普通高等教育"十一五"国家级规划教材

（高职高专教材）

工业分析专业英语

第三版

● 李居参　王佳欣　主编

● 李素荣　主审

化学工业出版社

·北京·

本书共分10部分（part）25单元（unit），选文广泛，难度适中，涵盖了分析化学、仪器分析、仪器使用、工业分析、有机分析、环境分析、油品分析、药物分析和食品分析等内容。同时，本书还在仪器的使用部分介绍了气相色谱、原子吸收分光光度计、酸度计和元素分析仪的使用及维护说明书等内容。本书第三版增加了可燃冰的仪器分析和润滑油的仪器分析内容。在科技英语基本技巧中特别增加英文摘要的写作技巧，供师生参考使用。

本书可供高职高专工业分析专业学生使用，也可供相关人员参考使用。

图书在版编目（CIP）数据

工业分析专业英语/李居参，王佳欣主编. —3 版.
北京：化学工业出版社，2018.8（2025.2 重印）
ISBN 978-7-122-32403-0

Ⅰ.①工… Ⅱ.①李…②王… Ⅲ.①工业分析-英语 Ⅳ.①TB4

中国版本图书馆 CIP 数据核字（2018）第 129906 号

责任编辑：蔡洪伟　陈有华　　　　　　　文字编辑：汲永臻
责任校对：王　静　　　　　　　　　　　装帧设计：王晓宇

出版发行：化学工业出版社（北京市东城区青年湖南街 13 号　邮政编码 100011）
印　　装：北京机工印刷厂有限公司
787mm×1092mm　1/16　印张 14¾　字数 361 千字　2025 年 2 月北京第 3 版第 6 次印刷

购书咨询：010-64518888　　　　　　　　售后服务：010-64518899
网　　址：http://www.cip.com.cn
凡购买本书，如有缺损质量问题，本社销售中心负责调换。

定　　价：39.00 元　　　　　　　　　　　　　　　　版权所有　违者必究

前　言

本教材自出版以来，在全国高职高专工业分析专业广泛使用，并得到教材使用院校的肯定和好评。广大师生在教材使用过程中，对本教材提出许多好的建议与要求。

近年来，随着科学技术的进步，新技术、新工艺、新领域、新设备的出现和发展，对工业分析专业毕业生专业英语的知识面和应用技能提出更高的要求。因此，本教材根据出版社的要求，依据教育部高职高专国家规划教材的建设精神，结合工业分析专业人才培养目标及实际教学的需求，组织相关专家、本课程授课教师、对应岗位工程技术人员对《工业分析专业英语》教材进行研讨与论证，充分吸收专家和一线师生的意见与建议。在第二版的基础上，本次修订增加了可燃冰的仪器分析和润滑油的仪器分析的内容。

本教材由李居参教授和大连工业大学王佳欣老师主编，宋凤军教授等参加了编写修订工作，由渤海大学李素荣教授主审。

本书的编写宗旨力求结构合理、内容丰富、体例清晰，反映了工业分析领域的新知识。由于教材内容涉及宽泛，难免出现疏漏，恳请读者给予斧正，使本书更趋臻善。

编　者

2018 年 4 月

第一版前言

本教材依据教育部高职高专教材的编写要求及高职高专工业分析专业教材编审委员会审定的高职高专《工业分析专业英语》教学基本要求进行编写，适用于高职高专工业分析及相关专业。

《工业分析专业英语》作为高职高专工业分析及相关专业人才培养的一门专业必修课程，其任务是通过本门课程的学习使学生能够掌握工业分析专业英语词汇及术语，掌握专业英语听、说、读、写、译五大技巧，具有正确、快速阅读英文科技文献的能力，初步具备专业英语的翻译及写作能力。同时通过该专业英语的学习，获取信息，进一步了解国内外工业分析专业的发展动态。

本书的编写宗旨是力求将专业知识和工业分析专业英语的学习与学生英语学习融为一体，既培养和提高学生阅读英语原著和报刊的能力，又向学生提供广泛的专业知识及有关的英语知识。全书共有20篇课文，20篇阅读材料，选材上注意与专业知识衔接；深度和广度适当，涉及分析专业诸多方面：包括分析化学、仪器分析、仪器使用、工业分析、有机分析、环境分析、油品分析、药品分析、食品分析等，均选自近年来英美等国的分析专业的期刊、文献；课文注释中解释了有关分析专业术语及背景知识；课文练习精炼实用；课后设有读、译、听、说、写五大科技英语基本技巧，书后附有专业术语词汇表和化学元素表，便于参考学习。

本书由辽宁石化职业技术学院李居参主编，天津渤海职业技术学院孙津清、辽宁石化职业技术学院司颐和刘亚珍分别编写了部分课文、阅读教材、练习题和注释。全书由李居参统稿，由渤海大学徐丽萍主审。

本教材涉及内容较广，可能出现错漏，希望读者不吝指正，使本书在使用过程中不断得到改进。

编 者
2004 年 1 月

第二版前言

本教材自 2004 年出版以来，在全国高职高专工业分析专业广泛使用，受到广大师生的首肯和欢迎，并被列入普通高等教育"十一五"国家级规划教材。随着新技术、新仪器的不断引进和发展，对工业分析专业人才专业英语的要求越来越高，特别对阅读仪器英文说明书方面的要求越来越高。本教材再版的主攻目标，旨在及时反映工业分析领域的最新技术，实现学生学习及岗位工作的对接。

为贯彻落实教育部有关高职高专教材建设精神，根据工业分析专业人才培养目标及本门课程实际教学需求，对《工业分析专业英语》第一版教材的部分内容进行补充和修订，在保留原有第一版教材内容和编写思路上，特别在仪器使用部分增加 3 个单元，精心选择了分析岗位使用较多的气相色谱、原子吸收分光光度计、酸度计和元素分析仪的使用及维护说明书等内容。

该教材由辽宁工业大学李居参教授主编，辽宁石化职业技术学院司颐和刘亚珍参加了编写修订工作。本教材由辽宁工业大学张祝祥教授主审。

我们的宗旨是通过修订给广大师生朋友们提供更好的切实的帮助和指导。由于时间仓促，疏漏之处在所难免，希望广大师生在使用本书的过程中，进一步提出宝贵意见，使本书更臻完善。

编 者
2008 年 3 月

Contents

PART 1　CHEMICAL ANALYSIS ································· 1
　Unit 1 Text：Titrimetric Analysis ································ 1
　　科技英语基本技巧　　翻译技巧：直译、意译 ··················· 4
　　　　　　　　　　　阅读技巧：快速阅读的四种方法 ··········· 5
　Reading Material　Qualitative Analysis Identifying Anions and Cations ········· 6
　Unit 2 Text：Acid-Base Titration ································· 9
　　科技英语基本技巧　　翻译技巧：科技术语的翻译方法 ········· 12
　　　　　　　　　　　阅读技巧：阅读速度检测方法 ············· 13
　Reading Material　Standardization of NaOH ····················· 14
　Unit 3 Text：Acid-Base Indicators ································ 16
　　科技英语基本技巧　　翻译技巧：词义选择法 ··················· 19
　　　　　　　　　　　阅读技巧：阅读中的不良习惯 ············· 19
　Reading Material　Analysis of a Sodium Carbonate and Sodium Bicarbonate
　　Mixture ··· 20
　Unit 4 Text：Precipitation Titration ······························ 22
　　科技英语基本技巧　　翻译技巧：英语形容词转译现象 ········· 25
　　　　　　　　　　　阅读技巧：快速阅读技巧 ··················· 26
　Reading Material　Mohr Method of Chloride Determination ······ 28
　Unit 5 Text：Redox Titration ····································· 30
　　科技英语基本技巧　　翻译技巧：英语介词的翻译 ············· 33
　　　　　　　　　　　阅读技巧1：眼睛是如何阅读的 ············ 36
　　　　　　　　　　　阅读技巧2：大脑后部的眼睛 ············· 36
　Reading Material　Determination of Hydrogen Peroxide（0.1%～6%）by
　　Iodometric Titration ·· 36
　Unit 6 Text：Complexometric Titration of Zn（Ⅱ）with EDTA ········· 39
　　科技英语基本技巧　　翻译技巧：英语数字的翻译 ············· 43
　　　　　　　　　　　阅读技巧1：较差阅读者 ··················· 46
　　　　　　　　　　　阅读技巧2：快速阅读者 ··················· 47
　Reading Material　Determination of Water Hardness by Complexometric
　　Titration ··· 47

PART 2　INSTRUMENT ANALYSIS ····························· 50
　Unit 7 Text：Determine Permanganate by Spectrophotometric Analysis ········ 50
　　科技英语基本技巧　　翻译技巧：被动语态的译法 ············· 54
　　　　　　　　　　　阅读技巧：快速阅读训练 ··················· 56
　Reading Material　Spectrophotometry ···························· 57

Unit 8 Text: Infrared Spectroscopy ·· 60
    科技英语基本技巧　翻译技巧：简单定语从句的汉译 ································· 64
                  听力技巧：语音方面的难点与策略 ································· 64
  Reading Material　Ultraviolet Spectroscopy ·· 65
Unit 9 Text: Potentiometric Titration ··· 68
    科技英语基本技巧　翻译技巧：状语从句的译法 ··· 71
                  听力技巧：语法方面的难点与策略 ···································· 72
  Reading Material　Principles of the Glass Electrode Method ················ 72
Unit 10 Text: Atomic Absorption Spectroscopy ··· 75
    科技英语基本技巧　翻译技巧：名词性从句的译法 ····································· 78
                  听力技巧：词汇方面的难点与策略 ···································· 79
  Reading Material　Atomic Absorption Spectrophotometer ···················· 80
Unit 11 Text: Gas Chromatography ·· 83
    科技英语基本技巧　听力技巧：听力与其他技能的关系 ································ 87
                  口语技巧：突破语音关 ·· 88
  Reading Material　High Performance Liquid Chromatography（HPLC）········· 88
Unit 12 Text: Mass Spectrometry ·· 93
    科技英语基本技巧　听力技巧：多听相关专业的英语授课或报告 ·················· 96
                  口语技巧：坚持"五说法" ·· 96
  Reading Material　Nuclear Magnetic Resonance Spectroscopy ············· 97

PART 3　INSTRUMENTS ·· 99
Unit 13 Text: Instruments ··· 99
    科技英语基本技巧　写作技巧：短文缩写阶段 ·· 103
                  听力技巧：篇幅方面的策略 ··· 104
  Reading Material　Science Lab Techniques ·· 105
Unit 14 Text: Spectrometer ··· 109
    科技英语基本技巧　科技英语写作句型：问题 ·· 112
                  听力技巧：题材方面的策略 ··· 113
  Reading Material　Single-Beam Spectrometer ··································· 114
Unit 15 Text: Clarus 400 GC Guide ·· 116
    科技英语基本技巧　写作技巧：科技论文英文摘要的类型 ······························ 122
  Reading Material　Clarus 400 GC Specification ·································· 123
Unit 16 Text: Optimizing PerkinElmer Flame AA ······································ 128
    科技英语基本技巧　写作技巧：科技论文英文摘要的内容 ······························ 133
  Reading Material　Features and Operation of Hollow Cathode Lamps ······· 135
Unit 17 Text: Operation of the Mettler S20 pH meter ······························ 139
    科技英语基本技巧　写作技巧：科技论文英文摘要的写作技巧 ······················· 142
  Reading Material　2410 Series Ⅱ Nitrogen Analyzer ························ 143

PART 4　INDUSTRY ANALYSIS ··· 147
Unit 18 Text: Sampling and Sub-sampling ··· 147

　　　　科技英语基本技巧　　科技英语写作句型：知识 ………………………………… 150
　　　　　　　　　　　　　口语技巧：专业性英语口头交流常用习语 ………………… 151
　　　Reading Material　Sample Preparation ………………………………………………… 151
PART 5　ORGANIC ANALYSIS ……………………………………………………………… 156
　　Unit 19 Text：Organic Compound Identification Using Infrared Spectroscopy ……… 156
　　　　科技英语基本技巧　　科技英语写作句型：研究 ………………………………… 161
　　　　　　　　　　　　　口语技巧：一般会话策略（Ⅰ） …………………………… 163
　　　Reading Material　Organic Analysis ……………………………………………………… 163
PART 6　ENVIRONMENTAL ANALYSIS ………………………………………………… 167
　　Unit 20 Text：COD is Determined by Potassium Dichromate …………………………… 167
　　　　科技英语基本技巧　　科技英语写作句型：方法 ………………………………… 171
　　　　　　　　　　　　　口语技巧：一般会话策略（Ⅱ） …………………………… 172
　　　Reading Material　Environmental Analysis ……………………………………………… 172
PART 7　OIL ANALYSIS …………………………………………………………………… 175
　　Unit 21 Text：Oil Analysis ………………………………………………………………… 175
　　　　科技英语基本技巧　　科技英语写作句型：实验 ………………………………… 179
　　　　　　　　　　　　　口语技巧：一般会话策略（Ⅲ） …………………………… 180
　　　Reading Material　Infrared Analysis as a Tool for Assessing Degradation in
　　　　　Used Engine Lubricants ……………………………………………………………… 180
PART 8　PHARMACEUTICAL ANALYSIS ……………………………………………… 183
　　Unit 22 Text：Spectrophotometric Analysis of Aspirin ………………………………… 183
　　　　科技英语基本技巧　　科技英语写作句型：理论 ………………………………… 186
　　　　　　　　　　　　　口语技巧：怎样讲得体的英语（Ⅰ） ……………………… 187
　　　Reading Material　Analytical Methods of Atropine …………………………………… 188
PART 9　FOOD ANALYSIS ………………………………………………………………… 191
　　Unit 23 Text：Determination of Nitrogen in Foodstuffs by the Kjeldahl Method …… 191
　　　　科技英语基本技巧　　科技英语写作句型：数据 ………………………………… 194
　　　　　　　　　　　　　口语技巧：怎样讲得体的英语（Ⅱ） ……………………… 195
　　　Reading Material　Determine Vitamin C in Fruit Juices ……………………………… 196
PART 10　INSTRUMENTAL ANALYSIS ………………………………………………… 198
　　Unit 24 Text：Instrumental Analysis of Gas Hydrates Properties ……………………… 198
　　　　科技英语基本技巧　　科技英语翻译技巧与实践 ………………………………… 201
　　　Reading Material　Gas Hydrates Structures and Promoters ………………………… 202
　　Unit 25 Text：Spectroscopic Analysis of Synthetic Lubricating Oil …………………… 205
　　　　科技英语基本技巧　　英文科技资料翻译的三大特点 …………………………… 208
　　　Reading Material　Spectroscopic Analysis of Synthetic Lubricating
　　　　　Oil-Results and Discussion ………………………………………………………… 210
Appendix Ⅰ　Glossary ……………………………………………………………………… 216
Appendix Ⅱ　A Table of Chemical Elements …………………………………………… 225

PART 1   CHEMICAL ANALYSIS

Unit 1

Text: Titrimetric Analysis

General Principles

Titrimetric analysis volumetrically measures the amount of reagent, often called a titrant, required to complete a chemical reaction with the analyte. A generic chemical reaction for titrimetric analysis is

$$a\text{A} + t\text{T} \longrightarrow \text{products}$$

Where a moles of analyte A contained in the sample reacts with t moles of the titrant T in the titrant solution.

The reaction is generally carried out in a flask containing the liquid or dissolved sample. Titrant solution is volumetrically delivered to the reaction flask using a burette. Delivery of the titrant is called a titration. The titration is complete when sufficient titrant has been added to react with all the analyte. This is called the equivalence point.

An indicator is often added to the reaction flask to signal when all the analyte has reacted. The titrant volume where the signal is generated is called the end point. The equivalence point and end point are rarely the same.

Successful Titrimetric Analysis

A few rules of thumb for designing a successful titration are:
- The titrant should either be a standard or should be standardized
- The reaction should proceed to a stable and well defined equivalence point
- The equivalence point must be able to be detected
- The titrant's and sample's volume or mass must be accurately known
- The reaction must proceed by a definite chemistry. There should be complicating side reactions
- The reaction should be nearly complete at the equivalence point. In other words, chemical equilibrium favors products
- The reaction rate should be fast enough to be practical

Types of Chemistry

Although any type of chemical reaction may be used for titrimetric analysis, the reactions most often used fall under the categories of

Acid-Base
$$HA + B \rightleftharpoons HB^+ + A^-$$

Oxidation-Reduction
$$Ox + Red \rightleftharpoons Red' + Ox'$$

Precipitation
$$M(aq) + nL(aq) \rightleftharpoons ML_n(s)$$

Complex Formation
$$M(aq) + nL(aq) \rightleftharpoons ML_n(aq)$$

Lewis Acid-Base chemistry is often involved in precipitation and complex.

Steps in a Titration

Titrimetric analysis generally involves the following steps:
- Sampling
- Titrant preparation
- Standard preparation and conversion to a measurable form
- Titrant standardization by titration of an accurately known quantity of standard
- Sample preparation and conversion to a measurable form
- Sample titration with the titrant solution
- Data analysis

Sample Calculations

Titrimetric analyses are often performed with single-point standardization. Standardization of a base titrant solution, used to determine the amount of acid in a sample, is often performed using a known, dry weight of potassium hydrogen phthalate (KHP) as the standard. There is one proton, or equivalent, per mole of KHP.

The titer value is determined from base titrant standardization. The titer value is the amount of acid neutralized per ml titrant solution.

Titration Curves

Titration curves are constructed by plotting $-\lg$ concentration, or pX, of the analyte versus the volume of titrant added to the reaction solution (see Figure 1-1). For example,

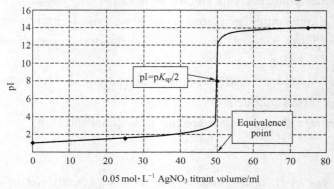

Figure 1-1 Titration curve for 25 ml 0.1 mol·L^{-1} NaI with 0.05 mol·L^{-1} AgNO$_3$ titrant

for acid-base titrations, the pH or pOH is plotted versus base or acid titrant volume. Titration curves may be plotted from experimental data to increase the precision of the resulting unknown determination. Similarly, theoretical titration curves may be used to investigate the feasibility of a titration.

Selected from "H. Diehl and G. F. Smith, Quantitative Analysis, John Wiley and Sons, Inc., New York."

New Words

titrimetric *adj.* [化] 滴定（测量）的
reagent *n.* 反应力，反应物，试剂
titrant *n.* [化] 滴定剂，滴定（用）标准溶液
flask *n.* 瓶，长颈瓶，细颈瓶，烧瓶，小水瓶，锥形瓶
burette *n.* [化] 滴定管，量管
equivalence *n.* 同等，[化] 等价，等值
indicator *n.* [化] 指示剂
stable *adj.* 稳定的
detect *vt.* 察觉，发觉，侦查，探测；*v.* 发现
volume *n.* 体积
mass *n.* 质量
category *n.* 种类，类别，[逻] 范畴
acid *n.* [化] 酸
base *n.* [化] 碱
oxidation *n.* [化] 氧化
reduction *n.* [化] 还原
precipitation *n.* 沉淀
sample preparation 样品制备
data *n.* 数据
calculation *n.* 计算，考虑
potassium *n.* [化] 钾（元素符号 K）
proton *n.* [核] 质子
curve *n.* 曲线，弯曲，[棒球] 曲线球，[统] 曲线图表
plotting *n.* 测绘，标图
experimental data 实验数据
precision *n.* 精确，精密度，精度

Notes

1. Titrimetric analysis volumetrically measures the amount of reagent, often called a titrant, required to complete a chemical reaction with the analyte.
   参考译文：在滴定中，与被测物质发生完全化学反应的试剂通常称为滴定剂，滴定分析

就是定量测出所消耗滴定剂的体积。

2. The reaction is generally carried out in a flask containing the liquid or dissolved sample. Titrant solution is volumetrically delivered to the reaction flask using a burette. Delivery of the titrant is called a titration. The titration is complete when sufficient titrant has been added to react with all the analyte.

参考译文：反应通常是在装有液体或溶解试样的锥形瓶中进行的。滴定时，使用滴定管将滴定剂定量滴加到锥形瓶中。滴加滴定剂的过程叫做滴定。只有当滴加的试剂与被测物质反应完全时，滴定过程才完成。

3. Standardization of a base titrant solution, used to determine the amount of acid in a sample, is often performed using a known, dry weight of potassium hydrogen phthalate (KHP) as the standard.

参考译文：标准碱溶液可以用来测定试样中酸的含量。在标定该标准碱溶液时，通常选用定量干燥的邻苯二甲酸氢钾作为基准物质。

4. The titer value is the amount of acid neutralized per ml titrant solution.

参考译文：滴定度是指每毫升碱滴定剂相当于所中和酸的量。

Exercises

1. Put the following into Chinese.

   titrimetric analysis   acid-base
   equivalence point   oxidation-reduction
   end point   precipitation

2. Put the following into English.

   指示剂   滴定度
   取样   数据分析
   滴定曲线   精密度

3. Questions.

   (1) Can you explain the definition of titrimetric analysis in your own words with the generic chemical reaction for titrimetric analysis?
   (2) How many rules of thumb for designing a successful titration?
   (3) What are the most common types of chemistry in titrimetric analysis?
   (4) Generally speaking, how many steps does titrimetric analysis involve?
   (5) Is the equivalence point different from the end point?

Lab Key

科 技 英 语 基 本 技 巧

翻译技巧：直译、意译

这里所讲的翻译方法指的是，通过英语和汉语两种语言特征的对比，分析其异同，阐述表达原文的一般规律。英、汉两种语言的结构存在一些共性，所谓的"直译"——既忠实原文内容，又符合原文的结构形式。但如果这两种语言之间还有很多差别，如完全地直译，势必出现"英化汉语"，这时就需要"意译"——在忠于原文内容的基础上，摆脱原文结构的束缚，使译文符合汉语的规范。特别要注意的是："直译"不等于"死译"，"意译"不等于"乱译"。

1. "直译"与"死译"的比较

原文结构与汉语结构一致，直译即可。但如原文结构与汉语结构不一致，仍采取直译的方法就变成了死译。如

（1）In chemistry, we use formulae to represent compounds. Compounds are composed of elements.

在化学中我们用分子式来表示化合物。化合物是由元素组成的。（"直译"）

（2）Solar energy seems to offer more hope than any other source of energy.

如果我们按原文结构原封不动地译成汉语，译文就变成了"太阳能似乎提供更多的希望比其他任何能源"。显然，该译文既不忠实原文含义，又不符合汉语的表达习惯。因此，必须运用"词类转换""词序调整""成分转换"等手段使译文通顺。例（2）一句可意译为："太阳能似乎比其他能源更有前途。"

2. "意译"不等于"乱译"

只有在正确地理解了原文结构和意义的基础之上，才能运用正确的方法调整原文结构，用规范的汉语表达出来，这才真正地做到"直译"。大量的翻译实践表明，大量英文句子的汉语翻译都要采取"意译法"。如果把"意译"理解为凭主观臆断来理解原文，可以不分析原文结构，只看词语表面的意义，自己编造句子，势必会造成"乱译"。如

（1）Nitrogen forms about four fifths of the atmosphere.

氮约占大气的五分之四。（forms 可以"意译"为"占"不必译为"形成"）

（2）It is easy to compress a gas, it is just a matter of reducing the space between the molecules. Like a liquid has no shape, but unlike a liquid it will expand and fill any container it is put in.

气体是很容易压缩的，这正是压缩分子之间的距离的根据。气体和液体一样，但又不同于液体，气体膨胀时会充满任何盛放它的容器。（"乱译"）

例（2）中 matter 一词的"乱译"是由于主观臆想（词义表述错误）造成的。在这里，将 matter 一词翻译成了"根据"在物理学和化学中是讲不通的，译为"压缩气体也就是减少分子之间的距离"，两者是一回事。在这里，matter 一词应该译为"事情"或"问题"。译为将主句"it will expand"转换为时间状语从句，导致表述不符合原义，应与同为谓语动词的"fill"一起翻译，译为"气体会膨胀充满"。综上所述，原译文可改为"气体很容易压缩，那只不过是缩小分子之间的距离而已。气体和液体一样没有形状，但又和液体不同，因为气体会膨胀并充满任何盛放它的容器。"

就翻译方法而论，总的说来，就是"直译法和意译法相结合"。又可细分为"照译""词义引申""成分转译"等译法。

阅读技巧：快速阅读的四种方法

阅读是从书面语言材料中获取社会行为意义的实践活动和心理过程。为适应高速发展的信息时代，快速阅读（fast reading）训练是非常必要的。美国爱德华·B·弗莱博士和韦德·E·卡特勒博士都曾著书阐明快速阅读的重要性，并总结了一系列完整的快速阅读理

论技巧。如果我们有计划、有针对性地了解有关知识和技巧，系统地进行快速阅读训练，那么可望更有效地提高阅读理解能力。

快速阅读是泛读的重要分支，但阅读速度要求快得多，在阅读过程中不作具体语言分析，只求抓住中心，了解大意，现在就让我们来看看练习快速阅读的四种方法。

1. 快速泛读（Fast Extensive Reading）

平时要养成快速泛读的习惯。这里讲的泛读是指广泛阅读大量涉及不同领域的书籍，要求读得快、理解和掌握书中的主要内容就可以了。要确定一个明确的读书定额，定额要结合自己的实际，切实可行，可多可少。例如每天读 20 页，一个学期以 18 周计算，就可以读 21 本中等厚度的书（每本书约 120 页）。

2. 计时阅读（Timed Reading）

课余要养成计时阅读的习惯。计时阅读每次进行 5～10 min 即可，不宜太长。因为计时快速阅读，精力高度集中，时间一长，容易疲劳，精力分散反而乏味。阅读时先记下"起读时间"（starting time），阅读完毕，记下"止读时间"（finishing time），即可计算出本次阅读速度。随手记下，长期坚持，必定收到明显效果。

3. 略读（Skimming）

略读又称跳读（reading and skipping）或浏览（glancing），是一种专门的、非常实用的快速阅读技能。所谓略读，是指以尽可能快的速度阅读，如同从飞机上鸟瞰（bird's-eye view）地面上的明显标志一样，迅速获取文章大意或中心思想。换句话说，略读是要求读者有选择性地进行阅读，可跳过某些细节，以求抓住文章的大概，从而加快阅读速度。据统计，训练有素的略读者（skimmer）的阅读速度可以达到每分钟 3000～4000 个词。

4. 寻读（Scanning）

寻读又称查读，同略读一样，寻读也是一种快速阅读技巧。熟练的读者善于运用寻读获得具体信息，以提高阅读效率。

寻读是一种从大量的资料中迅速查找某一项具体事实或某一项特定信息，如人物、事件、时间、地点、数字等，而对其他无关部分则略去不读的快速阅读方法。运用这种方法，读者就能在最短的时间内掠过尽可能多的印刷材料，找到所需要的信息。例如，在车站寻找某次列车或汽车的运行时刻，在机场寻找某次班机的飞行时刻，在图书馆查找书刊的目录，在文献中查找某一日期、名字、数字或号码等，都可以运用这种方法。

作为一种快速寻找信息的阅读技巧，寻读既要求速度，又要求寻读的准确性。具体地说，寻读带有明确的目的性，有针对性地选择问题的答案。因此，可以把整段整段的文字直接映入大脑，不必字字句句过目。视线在印刷材料上掠过时，一旦发现有关的内容，就要稍作停留，将它记住或摘下，既保证寻读的速度，又做到准确无误，所以寻读技巧也很有实用价值。

Reading Material

Qualitative Analysis Identifying Anions and Cations

Qualitative analysis is used to separate and detect cations and anions in a sample sub-

PART 1   CHEMICAL ANALYSIS

stance. In an educational setting, it is generally true that the concentrations of the ions to be identified are all approximately 0.01 mol·L^{-1} in an aqueous solution. The "semimicro" level of qualitative analysis employs methods used to detect 1~2 mg of an ion in 5 ml of solution.

First, ions are removed in groups from the initial aqueous solution. After each group has been separated, then testing is conducted for the individual ions in each group. Here is a common grouping of cations.

Group I: Ag^+, Hg^{2+}, Pb^{2+}
Precipitated in 1 mol·L^{-1} HCl
Group II: Bi^{3+}, Cd^{2+}, Cu^{2+}, Hg^{2+}, (Pb^{2+}), Sb^{3+} and Sb^{5+}, Sn^{2+} and Sn^{4+}
Precipitated in 0.1 mol·L^{-1} H$_2$S solution at pH 0.5
Group III: Al^{3+}, (Cd^{2+}), Co^{2+}, Cr^{3+}, Fe^{2+} and Fe^{3+}, Mn^{2+}, Ni^{2+}, Zn^{2+}
Precipitated in 0.1 mol·L^{-1} H$_2$S solution at pH 9
Group IV: Ba^{2+}, Ca^{2+}, K^+, Mg^{2+}, Na^+, NH_4^+
Ba^{2+}, Ca^{2+}, and Mg^{2+} are precipitated in 0.2 mol·L^{-1} (NH$_4$)$_2$CO$_3$ solution at pH 10; the other ions are soluble.

Many reagents are used in qualitative analysis, but only a few are involved in nearly every group procedure. The four most commonly used reagents are 6 mol·L^{-1} HCl, 6 mol·L^{-1} HNO$_3$, 6 mol·L^{-1} NaOH, 6 mol·L^{-1} NH$_3$. Understanding the uses of the reagents is helpful when planning an analysis.

Common Qualitative Analysis Reagents

Among the most common reactions in qualitative analysis are those involving the formation or decomposition of complex ions and precipitation reactions (see Table 1-1). These reactions may be performed directly by adding the appropriate anion, or a reagent such as H$_2$S or NH$_3$ may dissociate in water to furnish the anion. Strong acid may be used to dissolve precipitates containing a basic anion. Ammonia or sodium hydroxide may be used to bring a solid into solution if the cation in the precipitate forms a stable complex with NH$_3$ or OH$^-$ (see Table 1-2).

Table 1-1   Common qualitative analysis reagents

Reagent	Effects
6 mol·L^{-1} HCl	Increases [H$^+$]
	Increases [Cl$^-$]
	Decreases [OH$^-$]
	Dissolves insoluble carbonates, chromates, hydroxides, some sulfates
	Destroys hydroxo and NH$_3$ complexes
	Precipitates insoluble chlorides
6 mol·L^{-1} HNO$_3$	Increases [H$^+$]
	Decreases [OH$^-$]
	Dissolves insoluble carbonates, chromates, and hydroxides
	Dissolves insoluble sulfides by oxidizing sulfide ion
	Destroys hydroxo and ammonia complexes
	Good oxidizing agent when hot

Continued table

Reagent	Effects
6 mol·L^{-1} NaOH	Increases [OH$^-$] Decreases [H$^+$] Forms hydroxo complexes Precipitates insoluble hydroxides
6 mol·L^{-1} NH$_3$	Increases [NH$_3$] Increases [OH$^-$] Decreases [H$^+$] Precipitates insoluble hydroxides Forms NH$_3$ complexes Forms a basic buffer with NH$_4^+$

Table 1-2   Complexes of Cations with NH$_3$ and OH$^-$

Cation	NH$_3$ Complex	OH$^-$ Complex	Cation	NH$_3$ Complex	OH$^-$ Complex
Ag$^+$	[Ag(NH$_3$)$_2$]$^+$	—	Pb^{2+}	—	[Pb(OH)$_3$]$^-$
Al^{3+}	—	[Al(OH)$_4$]$^-$	Sb^{3+}	—	[Sb(OH)$_4$]$^-$
Cd^{2+}	[Cd(NH$_3$)$_4$]$^{2+}$	—	Sn^{4+}	—	[Sn(OH)$_6$]$^{2-}$
Cu^{2+}	[Cu(NH$_3$)$_4$]$^{2+}$ (blue)	—	Zn^{2+}	[Zn(NH$_3$)$_4$]$^{2+}$	[Zn(OH)$_4$]$^{2-}$
Ni^{2+}	[Ni(NH$_3$)$_6$]$^{2+}$ (blue)	—			

A cation is usually present as a single principal species, which may be a complex ion, free ion, or precipitate. If the reaction goes to completion the principal species is a complex ion. The precipitate is the principal species if most of the precipitate remains undissolved. If a cation forms a stable complex, addition of a complexing agent at 1 mol·L^{-1} or greater generally will convert the free ion to complex ion.

The dissociation constant K_d can be used to determine the extent to which a cation is converted to a complex ion. The solubility product constant K_{sp} can be used to determine the fraction of cation remaining in a solution after precipitation. K_d and K_{sp} are both required to calculate the equilibrium constant for dissolving a precipitate in a complexing agent.

Selected from "Novotny, M. Analytical Chemistry, 1989, Vol. 60, pp. 500~510."

Unit 2

Text: Acid-Base Titration

Titration is an analytical method in which a standard solution is used to determine the concentration of another solution.

Principles in Titrimetry
- A reaction with known stoichiometry is performed
- One reactant has a precisely known concentration (standard)
- One reactant has an unknown concentration (analyte)
- The reaction has a large equilibrium constant, to ensure the reaction proceeds to completion
- One reactant is delivered to the other in a series of known amounts. The equivalence point (specified as a volume of reactant added——sometimes called the equivalence volume) of the titration is that point when the reaction mixture has the reactants present in stoichiometric equivalence.
- For a reaction with the stoichiometry as specified below, the relation between concentrations and volumes is given by equation 1.

$aA + bB \longrightarrow$ products (a moles of reactant A, b moles of reactant B)

$$b/a(M_A V_A) = M_B V_B \qquad \text{equation 1}$$

where M_A——molarity of solution A (standard);
        V_A——volume of solution A;
        M_B——molarity of solution B (unknown);
        V_B——volume of solution B.

- For a titration to be successful, the reaction mixture must show a noticeable physical change when the mixture reaches the equivalence point. The point at which the physical change is observed is called the end point.
- The difference between the endpoint and equivalence point is the titration error.

Acid-base Titration

Any solution for which the concentration is precisely known is called a standard solution.

An acid-base titration uses the fact that one can be "neutralized" with the other. In this neutralization reaction, the acid and the base will be combined to produce a salt plus water. When done correctly, the resulting solution will be "neutral"——neither acid nor base. In a titration, this is known as the end point. The change in the pH of the solution can be monitored using an indicator or pH meter. It is extremely important that the exact amounts of each solution used be known at the end point.

$$NaOH + HCl \longrightarrow NaCl + H_2O$$

This balanced equation indicates that one mole of sodium hydroxide will combine with one mole of hydrochloric acid to produce one mole of sodium chloride and one mole of water. The only variable is the concentration of the two solutions.

This Figure 2-1 represents the titration of 10 ml of $0.1 \text{ mol} \cdot L^{-1}$ HCl with $0.1 \text{ mol} \cdot L^{-1}$ NaOH.

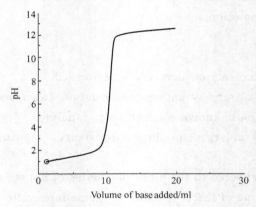

Figure 2-1　Titration of a strong acid with a strong base

The end point is characterized by a rapid change in pH with very little base added.

Phenolphthalein is the indicator we will use for an acid and base titration. Add two drops to the acid solution to be titrated. The solution will be colorless at this point. Slowly add the base solution until the faint pink color persists.

At the end point you know the concentration of the standard and the volume used, as well as the volume of the unknown used. Calculations can now be done to compare the number of moles of each solution used. This will give you the concentration of the unknown.

Selected from "Dorschel, C.A.; Ekmanis, J.L.; Oberholtzer, J.E.; Warren, F.V.Jr.; Bidlingmeyer, B.A. Analytical Chemistry, 1989, Vol.61, pp.951~968."

New Words

standard　*n.* 标准，规格，本位；*adj.* 标准的，第一流的
solution　*n.* 溶液
principle　*n.* 法则，原则，原理
titrimetry　*n.* ［化］滴定分析
stoichiometry　*n.* 化学计算（法），化学计量学
concentration　*n.* 集中，集合，专心，浓缩，浓度
equilibrium constant　平衡常数
ensure　*vt.* 保证，担保，使安全，保证得到；*vi.* 确保，保证
amount　*n.* 数量
molarity　*n.* ［化］［物］摩尔浓度

noticeable　*adj.* 显而易见的，值得注意的
titration error　滴定误差
neutralization　*n.* ［化］中和
produce　*vt.* 生成，生产，制造，结（果实）
plus　*prep.* 加上；*adj.* 正的，加的
neutral　*adj.* 中性的
monitor　*vt.* 监控；监测
meter　*n.* 米，计，表，仪表
sodium hydroxide　［化］氢氧化钠
hydrochloric acid　盐酸
chloride　*n.* ［化］氯化物
phenolphthalein　*n.* ［化］酚酞
colorless　*adj.* 无色的
faint　*adj.* 微弱的，暗淡的，模糊的
pink　*n.* 粉红色；*adj.* 粉红的

Notes

1. One reactant is delivered to the other in a series of known amounts. The equivalence point (specified as a volume of reactant added——sometimes called the equivalence volume) of the titration is that point when the reaction mixture has the reactants present in stoichiometric equivalence.

   参考译文：滴定分析是将一种已知浓度的标准溶液滴加到一定量被测物质中。所谓滴定化学计量点（指滴定完成时所滴加滴定剂的体积——有时也被称为等当体积）是指标准溶液与被测物质恰好按化学计量关系定量反应，达到化学计量点时所消耗的体积。

2. For a titration to be successful, the reaction mixture must show a noticeable physical change when the mixture reaches the equivalence point.

   参考译文：要准确地确定滴定终点，反应的混合物在达到化学计量点时必须表现出明显的物理变化。

3. This balanced equation indicates that one mole of sodium hydroxide will combined with one mole of hydrochloric acid to produce one mole of sodium chloride and one mole of water. The only variable is the concentration of the two solutions.

   参考译文：该平衡等式表明 1 mol 的氢氧化钠与 1 mol 的盐酸反应生成 1 mol 的氯化钠和 1 mol 的水。唯一的改变就是两种溶液的浓度。

4. At the end point you know the concentration of the standard and the volume used, as well as the volume of the unknown used.

   参考译文：在达到滴定终点时，可以得知标准溶液的浓度和消耗的体积以及所使用的未知溶液的体积。

Exercises

1. Put the following into Chinese.

equilibrium constant   acid-base titration
neutral   pH meter
balanced equation   hydrochloric acid
sodium chloride   phenolphthalein

2. Put the following into English.

标准溶液   滴定误差
中和   盐类
监测   氢氧化钠
浓度   体积

3. Questions.

(1) How can you define titration?
(2) What are the principles in titrimetry?
(3) Can you explain the equation "NaOH + HCl ⟶ NaCl + H$_2$O" in your own words?
(4) What is the end point?
(5) Is the end point different from the equivalence point?

Lab Key

科技英语基本技巧

翻译技巧：科技术语的翻译方法

随着社会的进步和科技的发展，新的发明创造不断涌现，随之也就出现了描述这些事物的新术语。在科技英语翻译中，我们常常碰到如何把这类术语译成适当的汉语的问题。通常有以下约定俗成的方法。

1. 意译

意译就是对原词所表达的具体事物和概念进行仔细推敲，以准确译出该词的科学概念。这种译法最为普遍，在可能的情况下，科技术语应采用意译法。例如

standard solution   标准溶液
titration error   滴定误差
stationary phase   固定相
equilibrium constant   平衡常数

一般来说，有三类科技新词语采用意译法。

(1) 合成词　由两个或两个以上旧词构成的复合词。

end point   终点
ground state   基态
white reference   基准白（色）
major repair   大修
buffer solution   缓冲溶液

(2) 多义词　旧词转义，即通过赋予旧词以新义而构成新术语。

concentration   集中，集合，专心→浓缩，浓度

（3）派生词　在原有的词根上加前缀或后缀构成新术语。
alkalinity　碱度
spectrophotometer　分光光度计
condenser　冷凝器
2. 音译
音译就是根据英语单词的发音译成读音与原词大致相同的汉字。采用音译的科技术语主要有两类。
（1）计量单位的词
mole　摩尔
decibel　分贝
bit　比特（度量信息的单位，二进制位）
joule　焦耳（功或能的单位）
（2）某些新发明的材料或产品的名称（尤其在起初时）
PVC　聚氯乙烯
Nylon　尼龙（聚酰胺树脂）
sonar　声呐（声波导航和测距设备）
Vaseline　凡士林（石油冻）
morphine　吗啡

一般来说，音译比意译容易，但不如意译能够明确地表达新术语的涵义。因此，有些音译词经过一段时间后又被意译词所取代，或者同时使用。
例如
El Nino　厄尔尼诺
Atropine　阿托品
Combine　康拜因→联合收割机
Aspirin　阿司匹林
Vitamin　维他命→维生素
Penicillin　盘尼西林→青霉素

阅读技巧：阅读速度检测方法

如何用最短的时间、最快的速度获取最丰富的信息，并对这些信息进行最有效的筛选，取得最佳效果，产生最好效应？对这些问题的圆满回答是：提高阅读速度，增强理解能力，改善记忆方式。

在进行系统的阅读技巧学习之前，首先应该对现阶段自身的阅读速度有所了解。具体检测方法是：所读文章单词数除以阅读所用的时间，计算出每分钟能阅读多少单词，即阅读速度（WPM）。

阅读速度计算公式：每分钟单词数（WPM）＝所读单词数/时间

阅读能力参考表

读者水平	阅读速度（WPM）	理解程度	读者水平	阅读速度（WPM）	理解程度
差	10～100	30%～50%	百里挑一	800～1000	80%以上
平均	200～240	50%～70%	千里挑一	1000以上	80%以上
专业人员	400	70%～80%			

Reading Material

Standardization of NaOH

Potassium Acid Phthalate

Potassium acid phthalate ($KHC_8H_4O_4$, or KHP) is one of the most common primary standards for sodium hydroxide titrant. The compound is a salt of phthalic acid ($H_2C_8H_4O_4$, or H_2P); only one proton per molecule of the potassium salt is acidic. The two molecules are represented:

$$H_2P \qquad KHP$$

(structures: phthalic acid with two COOH groups on benzene ring; KHP with one COOH and one COO^-K^+ on benzene ring)

Potassium acid phthalate reacts with sodium hydroxide according to the equation:

$$OH^- + HP^- \rightleftharpoons H_2O + P^{2-}$$
$$(NaOH) \quad (KHP)$$

The divalent phthalate ion is produced and is responsible for the basicity of the solution at the equivalent point. Phenolphthalein is employed as an indicator (acid form colorless, basic form red-pink; transition interval 8.0~9.6).

The same reaction is used for the determination of impure potassium acid phthalate unknowns, this is sometimes called the determination of the total acidity of a substance.

Preparation of $0.1\ mol \cdot L^{-1}$ Sodium Hydroxide

(1) Boil about 1 liter of distilled water for 5 minutes to remove carbon dioxide. Allow cooling covered with a watch glass; transfer while warm (40℃) to a one-liter bottle (bottles are stored in a hall cabinet).

(2) Transfer about 7 ml of a clear saturated (1:1) solution of sodium hydroxide to the bottle, using a transfer pipet and a rubber bulb. Mix thoroughly and keep covered.

(3) A saturated (1:1) solution of sodium hydroxide is usually available in the laboratory. Such a solution may be prepared by mixing 50g of sodium hydroxide pellets with 50 ml of water in a polyethylene bottle. Let it stand for a week to allow the isolable sodium carbonate to settle.

(4) Be sure to cool the polyethylene bottle when dissolving the sodium hydroxide. Use a water bath cooled with ice.

Standardization of $0.1\ mol \cdot L^{-1}$ Sodium Hydroxide

(1) Dry 4~5 g of primary-standard potassium acidic phthalate in a weighing bottle at 110℃ for 2 hours. Allow cooling 30 minutes in a desiccator before weighing (Why should the desiccator lid be left slightly "cracked" during the cool-down period?).

(2) Accurately weight out 800~900 mg of potassium acid phthalate into each of three

250ml flasks. This is about 4~4.5 and should require 40~50 ml of distilled water, warming if necessary. Cool before titrating.

(3) Review the steps on the use of the buret. Read buret to hundredths places, e. g. 10.43.

(4) Add 3 or more drops of phenolphthalein indicator to the first flask and titrate dropwise in the vicinity of the end point. If necessary, adjust the volume of phenolphthalein indicator added to the other two flasks, and titrate them too. Try to obtain the same intensity of pink color at the end point for all three titrations. Be sure to perform preliminary calculations in the laboratory. Repeat any experiment in which an unsatisfactory result is obtained.

(5) Calculate the normality of the sodium hydroxide by using an equivalent weight of 204.22 for potassium acid phthalate. Calculate relative average deviation. The precision should be 0.2%.

Selected from "Dorschel, C. A.; Ekmanis, J. L.; Oberholtzer, J. E.; Warren, F. V. Jr.; Bidlingmeyer, B. A. Analytical Chemistry, 1989, Vol. 61, pp. 951~968."

Unit 3

Text: Acid-Base Indicators

Introduction

Acid-base indicators (also known as pH indicators) are substances which change color with pH. They are usually weak acids or bases.

Consider an indicator, which is a weak acid, with the formula HIn. At equilibrium, the following chemical equation is established.

$$HIn(aq) + H_2O(l) \rightleftharpoons In^-(aq) + H_3O^+(aq)$$

  acid   base
  color A   color B

The acid and its conjugate base have different colors. At low pH, the concentration of H_3O^+ is high and so the equilibrium position lies to the left. The equilibrium solution has the color A. At high pH, the concentration of H_3O^+ is low and so the equilibrium position thus lies to the right and the equilibrium solution has color B.

Phenolphthalein is an example of an indicator, which establishes this type of equilibrium in aqueous solution.

colorless(acid)   magenta(base)

Phenolphthalein is a colorless, weak acid, which dissociates in water forming magenta anions. Under acidic conditions, the equilibrium is to the left, and the concentration of the anions is too low for the magenta color to be observed. However, under alkaline conditions, the equilibrium is to the right, and the concentration of the anion becomes sufficient for the magenta color to be observed.

We can apply equilibrium law to indicator equilibria in general for a weak acid indicator:

$$K_{In} = \frac{[H_3O^+][In^-]}{[HIn]}$$

K_{In} is known as the indicator dissociation constant. The color of the indicator turns from color A to color B or vice versa at its turning point. At this point:

$$[HIn] = [In^-]$$

The pH of the solution at its turning point is pK_{In} and is the pH at which half of the indicator is in its acid form and the other half in the form of its conjugate base.

Indicator Range

At a low pH, a weak acid indicator is almost entirely in the HIn form, the color of which predominates. As the pH increases, the intensity of the color of HIn decreases and the equilibrium is pushed to the right. Therefore, the intensity of the color of In$^-$ increases. An indicator is most effective if the color change is distinct and over a small pH range. For most indicators the range is within ± 1 of the pK_{In} value (see Table 3-1).

Table 3-1 Examples of common indicators

Indicator	Color		pK_{In}	pH range
	Acid	Base		
Thymol blue-1st change	red	yellow	1.5	1.2~2.8
Meth orange	red	yellow	3.7	3.2~4.4
Bromocresol green	yellow	blue	4.7	3.8~5.4
Meth red	yellow	red	5.1	4.8~6.0
Bromothymol blue	yellow	blue	7.0	6.0~7.6
Phenol red	yellow	red	7.9	6.8~8.4
Thymol blue-2nd change	yellow	blue	8.9	8.0~9.6
Phenolphthalein	colorless	magenta	9.4	8.2~10.0

Indicators are used in titration solutions to signal the completion of the acid-base reaction. A universal indicator is a mixture of indicators, which give a gradual change in color over a wide pH range. The pH of a solution can be approximately identified when a few drops of universal indicator are mixed with the solution.

Selected from "Wellinder, B.S.; Kornfelt, T.; Sorensen, H.H.; Analytical Chemistry, 1995, Vol. 67, pp. 39~43."

New Words

substance   *n.* 物质，实质，主旨
formula   *n.* 公式，规则，客套语，分子式
equilibrium   *n.* 平衡，平静，均衡
conjugate   *v.* 变化，共轭
aqueous   *adj.* 水的，水溶液的
magenta   *n.* 红紫色，洋红色
dissociate   *v.* 分离，游离，分裂
anion   *n.* 阴离子
alkaline   *adj.* [化] 碱的，碱性的
vice versa   反之亦然
predominate   *vt.* 掌握，控制，支配；*vi.* 统治，成为主流，支配，占优势
intensity   *n.* 强烈，剧烈，强度
thymol blue   百里酚蓝
methyl   *n.* 甲基，木精

bromocresol green　溴甲酚绿
phenol red　酚红
bromothymol blue　溴百里酚蓝
universal indicator　通用指示剂

Notes

1. Acid-base indicators (also known as pH indicators) are substances which change color with pH. They are usually weak acids or bases.

    参考译文：酸碱指示剂（亦称为 pH 指示剂）是指颜色随着 pH 的变化而发生变化的物质。它们通常为弱酸或弱碱。

2. Phenolphthalein is an example of an indicator, which establishes this type of equilibrium in aqueous solution.

    参考译文：酚酞是指示剂中的一种，它在水溶液中处于如下酸碱平衡状态。

3. Phenolphthalein is a colorless, weak acid, which dissociates in water forming magenta anions.

    参考译文：酚酞是一种无色的弱酸，当它溶解于水中时形成红紫色阴离子。

4. At a low pH, a weak acid indicator is almost entirely in the HIn form, the color of which predominates.

    参考译文：pH 较低时弱酸指示剂几乎完全呈酸色，酸色为主导颜色。

5. Indicators are used in titration solutions to signal the completion of the acid-base reaction.

    参考译文：指示剂加入滴定溶液中借助颜色变化作为酸碱滴定反应完全的信号。

Exercises

1. Put the following into Chinese.

    acid-base indicator　　　　　　　weak acid
    aqueous solution　　　　　　　　acidic condition
    equilibrium law　　　　　　　　　dissociation constant
    thymol blue　　　　　　　　　　　bromothymol blue

2. Put the following into English.

    弱碱　　　　　　　　　　　　　　碱性条件
    变色范围　　　　　　　　　　　　甲基红
    酚红　　　　　　　　　　　　　　通用指示剂
    化学平衡

3. Questions.

    (1) How can you define the term "acid-base indicator"?

    (2) Can you explain the equation $K_{In} = \dfrac{[H_3O^+][In^-]}{[HIn]}$ in simple English?

    (3) What are the common indicators?

    (4) Which kind of indicators is most effective?

    (5) What is a universal indicator?

科技英语基本技巧

翻译技巧：词义选择法

　　词是语言的基本单位。在英语和汉语这两种语言中，都存在着一词多用，一词多义的现象。因此翻译时应根据具体情况，选择符合译文要求的词义。一般可以从以下几个方面入手选择词义。

　　在英语和汉语中，同一个词在不同的学科领域或不同的专业中往往具有不同的词义。因此，在选择词义时，应考虑到阐述内容所涉及的概念是属于何种学科、何种专业。像英语名词 a carrier 在不同的学科、专业性场合中就具有不同的词义："搬运工"[交通]，"传导管"[运输]，"支架"[机械]，"托板"[机械]，"底盘"[车辆]，"载波"[通信]，"载流子"[半导体]，"载体"[化学]，"母舰"[军事]。根据上下文逻辑关系和语气连贯性来选择词义，这也是一种常用方法。

　　像英语词 develop
　　(1) Plants develop from seeds. 植物由种子发育而成。
　　(2) We must develop the natural resources of our country. 我们必须开发我们国家的自然资源。

　　像汉语词"也"
　　(1) 这里也很少见暴雨，所以用不着修溢洪道。Moreover, cloudburst are very uncommon in this area, therefore we don't have to construct safety outlets.
　　(2) 这种植物长在非洲，也长在南美洲。This crop grows in South America as well as in Africa.

　　在翻译时，还应注意根据搭配习惯来选择译文词义。像英语词 large：large land（广阔的土地），large farmers（大农场主），talk at large（详谈）。

　　像汉语词"坏"
　　(1) 这耕犁坏了。The plow is broken.
　　(2) 谷物坏了（变质了）。The corn has gone bad.
　　(3) 这片地被山洪毁坏了。The field has been destroyed by mountain torrents.

阅读技巧：阅读中的不良习惯

　　(1) 不紧不忙，慢中求准，"视幅"（eye span）小。
　　(2) 逐字阅读，一字不漏。一次只能看一个单词。这种看法之所以是错误的，是由于人们的凝视能力可以扩展，另外阅读不是理解单个的单词，而是整体意思。
　　(3) 指指点点、借用笔、手指、尺子指读。
　　(4) 碰见生词就查字典，反复回视。
　　(5) 边读边译。
　　(6) 有声朗读或心里默读（vocalization, sub-vocalization）。

> 为获取信息进行的阅读，应在适度紧张、全神贯注中进行。快速阅读既要节省时间，又要提高阅读效果，以上阅读障碍都应在训练中加以克服。

Reading Material

Analysis of a Sodium Carbonate and Sodium Bicarbonate Mixture

Objective

To analyze a mixture of sodium carbonate (Na_2CO_3), sodium bicarbonate ($NaHCO_3$) and the neutral component using titration; To determine the % composition of this mixture.

Introduction

Your task in this experiment will be to analyze a mixture of Na_2CO_3, $NaHCO_3$ and a neutral impurity using a volumetric analysis technique. The titration is performed in two steps. In the first step, the standardized HCl solution is used to titrate the mixture in the presence of phenolphthalein as an indicator, end point pH about 8. By the time the solution reaches this pH, only sodium carbonate will have reacted with the acid. Therefore, the number of moles of sodium carbonate can be easily calculated knowing that the reaction that takes place is described by the following net ionic equation.

$$CO_3^{2-}(aq) + H^+(aq) \longrightarrow HCO_3^-(aq) \qquad \text{Reaction 1}$$

Note that sodium ions, being the spectator ions, are not included.

The second step entails finishing the titration of the mixture resulting from step 1, using methyl orange as an indicator, end point: pH about 5. The bicarbonate reacts with hydrochloric acid forming carbonic acid, which decomposes to carbon dioxide and water.

$$HCO_3^-(aq) + H^+(aq) \longrightarrow H_2CO_3(aq) \qquad \text{Reaction 2}$$
$$H_2CO_3(aq) \longrightarrow CO_2(g) + H_2O(l)$$

Number of moles of sodium bicarbonate can be easily determined from the stoichiometry of Reaction 2 and titration data. However, this value represents the total number of moles of bicarbonate.

This includes the amount that was formed in the Reaction 1 as well as the amount of sodium bicarbonate originally present in the unknown mixture.

Procedure

Step 1

(1) Prepare the buret with standardized $0.1 \text{ mol} \cdot L^{-1}$ HCl solution (the regular: rinse with HCl, fill, remove air from the tip). Record the exact molarity of the acid.

(2) Obtain an unknown sample, record its identification code. Weigh out 0.20 g ($\pm$0.01 g) of the unknown sample and transfer it to the 125 ml Erlenmeyer flask. Add

about 15 ml of distilled water and swirl to dissolve the sample. The solution must be colorless. If the sample does not dissolve completely, don't worry, it will when you start titration.

(3) Add two (2) drops of phenolphthalein indicator. The solution will turn pink.

(4) Record the initial volume of acid in the buret. Titrate with the standard HCl until the solution is colorless. It is extremely important not to over titrate! If you do, make sure that a note about it is written in your observations. Do not discard this solution!

(5) Record the final volume of the HCl solution in the buret.

Step 2

(6) Add 2 drops of methyl orange indicator to the colorless solution resulting from the previous titration. The solution will turn yellow.

(7) Record the initial volume of acid in the buret. Titrate with the standard HCl until the solution turns light salmon pink (or light peach pink, but keep in mind that the color is really a faint idea of either shade).

(8) Record the final volume of the HCl solution in the buret.

(9) Dispose of the solution from the flask and clean the flask, rinsing very well with tap, then distilled water.

(10) Repeat step 1 and step 2 two more times using a new 0.20 g sample of the same unknown each time.

Selected from "Wellinder, B. S.; Kornfelt, T.; Sorensen, H. H.; Analytical Chemistry, 1995, Vol. 67, pp. 39~43."

Unit 4

Text: Precipitation Titration

Most precipitation titrations employ silver nitrate, an important and widely used volumetric reagent, as a precipitating agent because it reacts quickly. It is often used in the determination of anions such as chloride that precipitate as silver salts. Several indicators can be used for end point determination in precipitation titrations. One such class of indicators is adsorption indicators.

For precipitation titrations there are three main end point detection methods we will look at.

1. Light Scattering

A precipitate is formed as a product during a precipitation titration. A precipitate makes the solution appear cloudy. The density of this cloudy solution can be used to locate the end point of the precipitation titration. After the equivalence point, no added precipitate can be formed. All of titrand (the solution in the beaker to which reactant is added) has been exhausted and any added titrant (the solution added through the buret) remains unreacted. A light beam that is incident on the titration solution will be increasingly scattered as the titration reaction proceeds. After the equivalence point the scattering levels off. A plot of scattered light intensity as a function of added reactant will produce a curve from which we can obtain an end point. This method has lower end point precision but has high sensitivity.

2. Potentiometry

Potentiometry is based on direct measurement of the free concentration of one of the ions in the titration reaction solution with an electrochemical cell. This method allows generation of a complete titration curve. We will discuss potentiometric measurements in the section on electrochemistry.

3. Visual Indicators

There are three important visual indicator schemes for end point detection in precipitation titrations, the Mohr method, Fajans method, and Volhard method. We will discuss each.

Mohr method

The Mohr method is a titration scheme for analysis of anions that react with Ag^+ to form a precipitate. Ag^+ is added as the titrant. After the equivalence point, the excess Ag^+ reacts with chromate ion by the following reaction:

$$Ag^+ + X^- (\text{analyte}) \longrightarrow AgX(s)$$

$$2Ag^+ \text{(after eq. pt.)} + CrO_4^{2-} \longrightarrow Ag_2CrO_4(s)$$

The $Ag_2CrO_4(s)$ is an orange-yellow precipitate. When the color change is observed, the titration is ceased.

Fajans method

During a precipitation titration, the precipitate that is formed has excess surface charge. If the ion in the titrant that is participating in the titration reaction has a negative charge, the precipitate will have an excess negative charge prior to the equivalence point. For example, if $AgNO_3$ solution (titrant) is added through a buret to a solution containing Cl^- (titrand), the titration reaction is:

$$Ag^+ + Cl^- \longrightarrow AgCl(s)$$

Prior to the equivalence point, Cl^- is in excess in the reaction solution and the AgCl thus formed will have a small negative surface charge. After the equivalence point Ag^+ is in excess so the precipitate AgCl will have a positive surface charge.

In the Fajans method, a small amount of a colored indicator is added to the titration solution. The indicator is an adsorption indicator that adsorbs to the surface of the precipitate particles from the titration reaction. The indicator changes color if the precipitate surface changes. Thus, the solution visibly changes color from before to after the equivalence point.

Volhard method

The Volhard method is used for titrating a solution containing Ag^+. Other analytes are determined in coupled reaction schemes where the analyte reacts with excess Ag^+. The excess Ag^+ is then back-titrated with SCN^-. After the Ag^+/SCN^- equivalence point is reached, the excess SCN^- reacts with added Fe^{3+} (added in a small amount as an indicator) to form $FeSCN^{2+}$, which is a red complex.

$$Ag^+ \text{(in excess)} + X^- \text{(analyte)} \longrightarrow AgX(s)$$
$$Ag^+ \text{(remaining excess)} + SCN^- \longrightarrow AgSCN(s)$$
$$SCN^- \text{(excess after equivalence point)} + Fe^{3+} \longrightarrow [FeSCN]^{2+} \text{(red complex)}$$

At the first observation of the red color, the titration is ended and the titrant volume is noted. Again, your book has tabulated analytes that may be analyzed via this scheme.

Selected from "Synder, L. R.; Stadalius, M. A.; Quarry, M. A. Analytical Chemistry, 1983, Vol. 55, pp. 1412~1430."

New Words

employ   *vt.* 雇用, 用, 使用; *v.* 使用
silver nitrate   [化] 硝酸银
precipitating agent   沉淀剂
adsorption   *n.* 吸附
detection   *n.* 察觉, 发觉, 侦查, 探测, 发现
beaker   *n.* 大口杯, 有倾口的烧杯
exhausted   *adj.* 耗尽的, 疲惫的

beam   *n*. 梁，桁条，（光线的）束，柱，电波，横梁
scatter   *v*. 分散，散开，撒开，驱散
curve   *n*. 曲线，弯曲，[棒球]曲线球，[统]曲线图表
sensitivity   *n*. 敏感，灵敏（度），灵敏性
potentiometry   电势[位]测定法
direct measurement   直接测量
ion   *n*. 离子
electrochemistry   *n*. [化]电化学
Mohr method   莫尔法
Fajans method   法扬司法
Volhard method   福尔哈德法
chromate   *n*. 铬酸盐
charge   *n*. 负荷，电荷
positive   *adj*. 正的，阳的
colored indicator   显色指示剂
adsorption indicator   吸附指示剂

Notes

1. Most precipitation titrations employ silver nitrate, an important and widely used volumetric reagent, as a precipitating agent because it reacts quickly.

    参考译文：硝酸银是一种应用广泛的重要的容量分析试剂。由于它反应速率较快，所以大多数沉淀滴定都采用硝酸银作为沉淀剂。

2. Several indicators can be used for end point determination in precipitation titrations. One such class of indicators is adsorption indicators.

    参考译文：在沉淀滴定中有几种指示剂可以用来确定滴定终点。其中的一类就是吸附指示剂。

3. If the ion in the titrant that is participating in the titration reaction has a negative charge, the precipitate will have an excess negative charge prior to the equivalence point.

    参考译文：如果在滴定反应过程中参加反应的滴定剂离子带有负电荷，那么在到达化学计量点之前沉淀物就会有一个游离的负电荷。

4. In the Fajans method, a small amount of a colored indicator is added to the titration solution. The indicator is an adsorption indicator that adsorbs to the surface of the precipitate particles from the titration reaction. The indicator changes color if the precipitate surface changes. Thus, the solution visibly changes color from before to after the equivalence point.

    参考译文：在法扬司法中，将少量的显色指示剂滴加到滴定液中。该指示剂是一种吸附指示剂，可以吸附在滴定反应所形成的沉淀胶粒的表面上。如果沉淀物质的表面发生变化，该指示剂的颜色发生变化。因而，该溶液在化学计量点前后的颜色会发生明显的变化。

5. At the first observation of the red color, the titration is ended and the titrant volume is noted.

参考译文：一见到红色出现时，滴定就结束，同时记下滴定剂所消耗的体积。

Exercises

1. Put the following into Chinese.

   silver salts   adsorption indicator
   sensitivity   direct measurement
   electrochemistry   titration curve
   negative charge

2. Put the following into English.

   硝酸银   沉淀剂
   精密度   铬酸盐
   返滴定   吸附指示剂

3. Questions.

   (1) What is the most common used method of precipitation titration?

   (2) How many main end point detection methods do we have when we are performing the titration?

   (3) How do you explain the term "the Mohr method" in simple English?

   (4) Can you give the reason why we use silver salts in a precipitation titration most of the time?

   (5) How do you know that the titration is ended?

Lab Key

科 技 英 语 基 本 技 巧

翻译技巧：英语形容词转译现象

英语和汉语语言结构和表达习惯有很多差异之处，翻译时不能按原文逐词逐句死译。现介绍形容词的转译现象。

1. 在很多情况下，将英语的"形容词＋名词短语"译成汉语的主谓结构，译文显得顺口。

(1) In order to get a large amount of water power we need a large pressure and a large current. 为了获得大量的水力，我们需要高的水压和强的水流。

(2) An insulator offers a very high resistance to the passage through it of electric current. 绝缘体对电流通过有很大的阻力。

2. 英语中一些形容词连同系动词构成复合谓语时，翻译时可将形容词译成动词。

(1) The method is free from interference by other volatile material or organic matter. 此法免除了其他的挥发物或有机物的干扰。

(2) Salt and sugar are both soluble in water. 盐和糖都能溶于水。

3. 为了便于表达，一些英语词可译为汉语名词。

（1）Contamination leads to lower yield. 污染导致低产。

（2）After a few months, our just-in-time system became so efficient. 几个月后，我们的"准时供货"办法变得很有成效。

4. 由于语言习惯不同，英语里的形容词有时译成汉语副词。

（1）For efficient, safe and smoke-free operation, the distribution, quantity, temperature and velocity of the secondary air must be adequate to insure proper mixing and temperature. 为了有效、安全、无烟地操作，二次空气的分布、数量、温度和速度必须足以保证适当的混合和温度。

（2）Heat of reaction ample to maintain reaction can be evolved during smelting sulfide copper concentrates. 在熔炼硫化精矿时能放出足以维持反应的反应热。

阅读技巧：快速阅读技巧

1. 阅读文学材料的快速性

要想达到快速阅读的目的，关键是眼部机能的训练，即用特殊方法使眼部机能灵活自如，达到视角、视幅、视停、视移等视觉最佳状态，使视线如行云流水般地快速阅读。训练方法可按手指法（即目光随着手指左右，上下移动，头不要摇动）、图谱法（如点、圆、抛物线等图形目光沿着图形而快速移动）、词谱法、投影仪进行快速阅读的基本功训练等。当眼部机能训练适应之后，可采用快速阅读初级方法之一——跳读法，所谓跳读法就是指眼光从一"字群"跳到另一个"字群"进行识读（字群是由多个单词组成的）。这个过程眼球按"凝视—跳跃—凝视"的程序进行连续，不断运动，如

The man in/the brown coat/was reading a book.

当跳读练习熟练之后，可进行练习扩大视力识读文字的单位面积的训练。首先进行5个单词的练习，练习是主视区总应放在中间，也就是主视中间的3个单词，两边单词用余视力扫视。如

/We/have a colour/TV/.

在练习5个单词达到熟练之后，可加宽视区练习，一下看6个单词，7个单词，甚至达到9个单词，逐渐加宽视区范围，延长目光移视长度，这样就能缩短凝视时间，达到快速阅读的目的。

2. 阅读文字材料的无声性

上面我们介绍的只是快速阅读的先决条件，速读的关键还在于"无声"训练，在阅读速度上，无声要比有声快，这是因为有声阅读是眼、脑、口、耳四个器官一起活动，文字符号反映到眼睛，再传到大脑，大脑命令嘴发声，耳在监听辨别正确与否。而无声阅读只是运用眼和脑两大器官，省去了口的发声和耳朵的监听，因而它的速度要快。

快速阅读的信息变换方式为：书面信息→眼睛扫描信息→大脑记忆中枢的信息。因此我们应用特殊的方法和手段消除读声和心声，特殊手段就是用自身单声调鼻音，单声调心声或外界背景音乐抵消并消除读声和心声的手段，对特殊顽固的不发声不能阅读的人，还

可用一套自创歌曲，边唱边读，最后达到无声阅读。

3. 阅读方法的科学性

我们在阅读的时候，必须通过直觉、联想、想象、逻辑分析和综合判断等一系列思维活动，才能把顺次进入视觉的一连串文字信号转换成概念和思想，完成阅读过程，要完成其过程，必须进行科学阅读，进行科学阅读应具备以下几个条件。

（1）自信心　一个人要想在快速阅读上获得成功，首先要有自信心，在快速阅读时，自信心是很重要的，只要我们坚信我们能成功，通过长期苦练就会实现的。

（2）集中注意力　快速阅读的同时还要求快速记忆，这就要求在阅读时，不仅要阅读，而且要记、要理解，这是一个高难度的思维活动，没有集中的注意力很难保证"速读"的完成。

（3）快速理解——快速阅读的催化剂　"理解"就是利用已有的知识经验，去获得新的知识经验，并把新的知识经验纳入已有的知识经验系统中，理解可分为直接理解和间接理解。直接理解就是在瞬息之间立刻实现的，不需要任何中间思维过程，与知觉融合在一起。在这种情况下，主要是通过瞬间忆起以前所得的知识，选取立刻所需要的知识。而间接理解的实现需要通过一系列复杂的分析综合活动。快速阅读用的是中间理解法，它包括：推断法，即实行快速阅读的人往往根据几个单词推断出一个句子，由句子推知整个段落的意思，这就需要多读书，知识积累越多，知识面越宽，理解力越强，快速阅读中的推断能力才能越高。正是由于这种推断，眼睛才能停顿到最有信息含义的地方上。英语中使用的推断法之一是学会略过那些无关紧要的词汇。

如：The usual life span for Shanghai men is 72 years. 如果我们阅读时不知道"span"的词义，我们也完全可以看懂句子意思是"通常上海男子的寿命是72岁"。推断法之二是利用英语构词法推断词义，构词法由转换、派生与合成三部分构成。

（4）抓住关键词句　为了提高阅读速度首先应抓住关键词句，因为它是连接上下文的纽带，快速阅读时只注意瞬时关键词，其他便可迎刃而解，抓住关键句子也就是找出主题句，主题句是文章中用来概括大意的句子，主题句往往是每个段落的第一个句子，有时可能是最后一个句子，在特殊情况下可能出现在段落当中，通过识别主题句，可以快速、准确地抓住文章中各个段落的主要意思，如果能把每一段落的大意抓住了，那么全篇文章的中心思想也就把握住了，在阅读中识别主题句，并准确理解其意思，可帮助我们了解作者的行文思路，分析文章的内容结构，搞清楚各个段落之间的逻辑关系，有利于提高阅读的速度和理解的准确性。

（5）快速阅读能促使快速记忆　快速阅读时人的注意力高度集中，连续的快速阅读是一种强化活动，强化活动能够巩固和促进快速记忆的成果。强化记忆有三个层次：一是死记硬背（这是必要的、不可缺少的层次）；二是联想记忆；三是理解记忆。以阅读现代记叙文类（童话故事，作文选等）为例：要求硬记的是文题、作者、文中时间、地点、人物、姓名、名人名句等；要求联想记忆的故事情节（事件起因，事件发展，关键情节，高潮情节，事件结局）；要求理解记忆的关键词、关键句、中心语、段首语、事件性质、人物命运、作者态度、人称变化、词语概念、文章含义或中心思想等。总之，快速阅读能促进理解的质量、理解的速度和快速记忆。

Mohr Method of Chloride Determination

In a precipitation titration, the stoichiometric reaction is a reaction, which produces in solution a slightly soluble salt that precipitates out. To determine the concentration of chloride ion in a particular solution, one could titrate this solution with a solution of a silver salt, say silver nitrate, whose concentration is known. The chemical reaction occurring is $Ag^+(aq) + Cl^-(aq) \longrightarrow AgCl(s)$. A white precipitate of AgCl is deposited on the bottom of the flask during the course of the titration. Since the chemical reaction is one silver ion to one chloride ion, we know that the amount of silver ion used to the equivalence point equals the amount of chloride ion originally present. Since by the definition of molarity $n = cV$, the number of moles of either silver ion or chloride ion can be calculated from the number of moles of the other, and the molar concentration or the volume of added solution can be calculated for either ion if the other is known.

The substance to be titrated is generally measured into the titration vessel either directly, its mass (or density and volume) having been determined, or by pipet if it is in the form of a solution. The titrant solution is generally delivered from a buret. The volume added can be measured from the buret scale as soon as the endpoint of the titration is reached.

P Notation

It is inconvenient to the point of being impractical to plot, or even to compare, the changes in ionic concentrations which take place over the course of a precipitation titration because the values of the concentrations cover so many orders of magnitude in range. Chemists have therefore introduced p notation, in which the negative logarithm of a concentration or activity is used rather than the concentration or activity itself; that is, $pX = -\lg c(X)$ or $pX = -\lg a(X)$. The logarithmic p notation is commonly used not only in titrations, but for the general expression of solution concentrations. In other section this notation, in the form of pH, is extensively used to express the acidity of solutions.

In the following Figure 4-1, pCl is used on the vertical axis to show how the concentra-

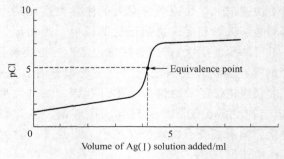

Figure 4-1 Titration curve

tion of chloride ion changes over the course of the titration. The molar concentration of either chloride ion or silver ion will change over several orders of magnitude during the course of this titration, as the concentration of chloride ion is first slowly reduced by the precipitation of AgCl as a consequence of the continuous addition of silver ion. As the supply of chloride ion is reduced to very low values, the equivalence point of the titration is reached——the point at which the stoichiometric precipitation is complete and the amount of silver ion that has been added is equivalent to the amount of chloride ion originally present. The term "equivalent" is used rather than "equal" because in some reactions, such as the precipitation of Ag_2SO_4, the amounts will differ by a stoichiometric factor of two or three. Beyond the equivalence point, addition of more silver ion will continue to reduce the concentration of chloride ion through the common ion effect.

Endpoints

In any titration, it is necessary to have some method of detecting when just enough of the titrant has been added——a procedure known as detecting the end point of the titration. The end point of this titration can be detected if the rapid change in either the concentration of silver ion or the concentration of chloride ion which occurs at the end point can be made apparent to an observer. Either instrumental methods or equilibrium methods can be used. The equilibrium methods are fairly straightforward. In this case we can use Ag_2CrO_4, because a solution of CrO_4^{2-} is yellow while a solution or precipitate of Ag_2CrO_4 is blood-red.

Suppose $[CrO_4^{2-}]$ in the solution is 0.001 molar. Then $K_{sp} = 1.12 \times 10^{-12} = [Ag^+]^2 [CrO_4^{2-}]$. Since $[CrO_4^{2-}] = 10^{-3}$ molar, $[Ag^+]^2 = 1.12 \times 10^{-10}$, and $[Ag^+] = 3.35 \times 10^{-5}$. At any concentration of silver ion greater than 3.35×10^{-5} molar in such a solution, a precipitate of Ag_2CrO_4 will form. If the concentration is below this, no silver chromate precipitate will form.

When we have a solution of, say, 0.01 molar chloride ion and add silver ion to it, the solubility product is $K_{sp} = 1.76 \times 10^{-10} = [Ag^+][Cl^-]$, so at the start of the titration $[Ag^+] = 1.76 \times 10^{-10} / 1 \times 10^{-2} = 1.76 \times 10^{-8}$ and no precipitate of Ag_2CrO_4 can form. At the equivalence point, $[Ag^+] = [Cl^-]$ and $[Ag^+]^2 = K_{sp} = 1.76 \times 10^{-1}$, $[Ag^+] = 1.33 \times 10^{-5}$ and no precipitate will form. But when a drop or two more of silver nitrate solution is added after the equivalence point has been reached, there is no more chloride ion to react with it. The concentration of silver ion may go up to say 10^{-3} molar. The solubility product of silver chromate will then be exceeded and a red precipitate of Ag_2CrO_4 will designate the end of the titration.

Selected from "Synder, L.R.; Stadalius, M.A.; Quarry, M.A. Analytical Chemistry, 1983, Vol. 55, pp. 1412~1430."

Unit 5

Text: Redox Titration

Introductory Theory

Oxidation is the name given to a chemical process in which an atom or ion loses one or more electrons. Reduction is the name given to a chemical process in which an atom or ion gains one or more electrons. Obviously, in order for a species to gain an electron, some other species has to lose an electron. Therefore, oxidation and reduction always go together. In fact, the two combined processes are often known as "redox".

Substances, which are reduced, do so by causing other objects to be oxidized, so they are often referred to as oxidizing agents. Substances, which are oxidized, do so by causing other objects to be reduced, so they are known as reducing agents.

In this experiment, a redox reaction will be performed in order to help us do a quantitative analysis. We have a solution of $KMnO_4$, a strong oxidizing agent. We know that its concentration is approximately $0.01 \ mol \cdot L^{-1}$, but we cannot prepare a solution of it whose concentration is accurately known. This is because the $KMnO_4$ will react with small amounts of organic material present in the water when the solution is first prepared. As a result, the concentration will never be as high as calculated. We are going to react a known volume of $KMnO_4$ with a known volume of a reducing agent $FeSO_4(NH_4)_2SO_4 \cdot 6H_2O$. The concentration of this solution can be accurately known. We will perform stoichiometric calculations to determine the concentration of the potassium permanganate. The combination of the two solutions gives the net ionic equation:

$$5Fe^{2+} + MnO_4^- + 8H^+ \longrightarrow Mn^{2+} + 5Fe^{3+} + 4H_2O$$

Unlike acid-base titration, where an indicator is required, this reaction is self-indicating. The permanganate ions give their solution a very dark purple color. As they react with the ferrous ions, the purple disappears. If even a fraction of a drop of permanganate is added after the equivalence point (the point when just enough permanganate has been added to react with all of the iron present), there will be no ferrous ions to remove the color, and there will be a persistent pink color. It is critical to titrate only until the faintest persistent pink color can be observed.

Procedure

(1) 100 ml of $0.0500 \ mol \cdot L^{-1} \ Fe^{2+}$ solution was prepared using 40 ml of $3 \ mol \cdot L^{-1}$ H_2SO_4 and 1.96 g of the iron (Ⅱ) ammonium sulfate hexahydrate.

(2) The titration apparatus was prepared and the burette was filled with $KMnO_4$ of unknown concentration. 10 ml of the $0.0500 \ mol \cdot L^{-1} \ Fe^{2+}$ solution was put into a flask.

(3) Three titrations of the unknown $KMnO_4$ solution into 10ml of the Fe^{2+} solution were performed and the data was recorded. The end point occurs when a very faint pink color persists.

(4) Watch to make sure that you do not go over the end point. Even one drop too much will make the solution a very dramatic pink color (see Table 5-1).

Table 5-1   Experiment data

Project	Trial 1	Trial 2	Trial 3
Volume of Fe^{2+} solution/ml	10.0	10.0	10.0
Initial burette reading/ml	15.82	31.08	49.39
Final burette reading/ml	5.05	21.52	38.10
Volume of $KMnO_4$/ml	10.77	9.56	10.29
Concentration of $KMnO_4$/mol·L^{-1}	0.009285	0.010500	0.009718

Data Table

Analysis-Sample Calculations

$$5Fe^{2+} + MnO_4^- + 8H^+ \longrightarrow Mn^{2+} + 5Fe^{3+} + 4H_2O$$

0.0500 mol·L^{-1} x mol·L^{-1}

10.0 ml   10.77 ml

0.0500 mol·L^{-1} Fe^{2+} = (0.0500 mol Fe^{2+}/L) × 0.0100 L × (1 mol MnO_4^-/5 mol Fe^{2+}) × (1/0.01077L) = 0.00929 mol·L^{-1} MnO_4^-

Average $KMnO_4$ concentrations: 0.00984 mol·L^{-1}

Conclusion

The average molar concentration of the aqueous potassium permanganate solutions used in the lab's three trials was 0.00984 mol·L^{-1}.

Selected from "Haugland, R.P., Handbook of Fluorescent Probes and Research Chemicals; Molecular Probes Inc.; Eugene, OR, 1985."

New Words

oxidation   *n*. [化] 氧化
electron   *n*. 电子
reduction   *n*. [化] 还原
redox   *n*. 氧化还原作用
oxidize   *v*. (使)氧化
quantitative   *adj*. 定量的，数量的
analysis   *n*. 分析，分解
approximately   *adv*. 近似地，大约
accurately   *adv*. 正确地，精确地
organic   *adj*. 有机的
potassium permanganate   [化] 高锰酸钾

ionic *adj*. 离子的
purple *adj*. 紫色的；*n*. 紫色
ferrous *adj*. 铁的，含铁的，[化]亚铁的
persistent *adj*. 持久稳固的
ammonium *n*. [化]铵
sulfate *n*. [化]硫酸盐
hexahydrate *n*. [化]六水合物
apparatus *n*. 器械，设备，仪器
flask *n*. 瓶，长颈瓶，细颈瓶，烧瓶，小水瓶
initial *adj*. 最初的，词首的，初始的
average *n*. 平均，平均水平，平均数

Notes

1. Obviously, in order for a species to gain an electron, some other species has to lose an electron. Therefore, oxidation and reduction always go together. In fact, the two combined processes are often known as "redox".

    参考译文：很显然，若使一个原子或离子得到一个电子，那么另外一个原子或离子就必须失去一个电子。因此，氧化和还原总是同时进行。实际上，两个过程通常合称为"氧化还原"。

2. This is because the $KMnO_4$ will react with small amounts of organic material present in the water when the solution is first prepared. As a result, the concentration will never be as high as calculated.

    参考译文：这是因为最初配制溶液时，高锰酸钾会和水中少量有机物发生反应。结果，溶液浓度永远达不到计算值那么高。

3. Unlike acid-base titration, where an indicator is required, this reaction is self-indicating. The permanganate ions give their solution a very dark purple color. As they react with the ferrous ions, the purple disappears.

    参考译文：酸碱滴定中需要使用指示剂，而氧化还原反应和酸碱滴定不同，它是靠自身来指示终点的。高锰酸根离子会使溶液呈深紫色。当高锰酸根离子与亚铁离子反应时紫色便消失。

4. We know that its concentration is approximately $0.01 \text{ mol} \cdot L^{-1}$, but we cannot prepare a solution of it whose concentration is accurately known.

    参考译文：我们知道溶液浓度大约是 $0.01 \text{ mol} \cdot L^{-1}$，但是我们配制不出准确浓度的高锰酸钾溶液。

5. If even a fraction of a drop of permanganate is added after the equivalence point (the point when just enough permanganate has been added to react with all of the iron present), there will be no ferrous ions to remove the color, and there will be a persistent pink color. It is critical to titrate only until the faintest persistent pink color can be observed.

    参考译文：如果在达到化学计量点（该点是指滴加足够的高锰酸钾与水中所有的铁离子恰好反应完全）之后，即使滴加一小滴高锰酸钾溶液，也不会有亚铁离子使溶液退色而且溶液会始终呈现粉红色。连续滴定直到可以观察到溶液呈现淡淡的粉红色时才结束滴定，这一点非常重要。

Exercises

1. Put the following into Chinese.

   reduction   quantitative analysis

   strong oxidizing agent   organic material

   potassium permanganate   self-indicating

   iron（Ⅱ）ammonium sulfate hexahydrate

2. Put the following into English.

   氧化   氧化还原反应

   亚铁离子   摩尔浓度

   平均浓度   硫酸铵

3. Questions.

   (1) What is the definition of "oxidation"?

   (2) Can you explain the term "reduction" in simple English?

   (3) Is acid-base titration the same with redox titration and why?

   (4) How many steps do you have to follow to perform a redox titration?

   (5) Can you illustrate the data table in text in English?

科 技 英 语 基 本 技 巧

翻译技巧：英语介词的翻译

英语中大多数介词含义灵活，一词多义多用。除了一些常用短语已有译法外，大量介词需要从其基本意义出发，联系上下文加以灵活处理。下面简明地介绍几种基本译法。

1. 转译

英语中常用介词来表达动作意义。汉译时，可将介词转译成动词。

(1) 在作表语的介词短语中，介词常转译为动词，而系动词则省略不译。如

Copper and gold were available long before man has discovered the way of getting metal from compound. 在人们找到从化合物中提取金属的方法以前就使用铜和金了。（available 转译成动词"使用"）

(2) 在作目的或原因状语的介词短语中，介词有时转译成动词。如

Kerosene is not so volatile as gasoline. 煤油不像汽油那样容易挥发。

The letter E is commonly used for electromotive force. 通常用 E 这个字母表示电动势。

(3) 在作条件、方式或方法状语的介词短语中，介词有时转译成动词。如

But even the larger molecules with several hundred atoms are too small to be seen with the best optical microscope. 但是，即使有几百个原子的分子也是太小了，用最好的光学显

微镜也看不见它们。

（4）介词短语作补足语时，其中介词常转译成动词。如

Heat sets these particles in random motion. 热量使这些粒子做随机运动。

2. 增译

增词不是无中生有，而是要根据上下文特别是与介词搭配的动词或形容词的含义加得恰当。有不少情况，句中与介词搭配的动词或形容词不出现，如照原文结构无法把意思表达清楚，甚至易于误解时，这就需要增词。如

In rapid oxidation a flame is produced. 在快速氧化的过程中会产生火焰。（对原句中 oxidation 增译为氧化过程）

Resonance is often observed in nature. 在自然界当中常常观察到共振现象。（Resonance 由原来的"共振"增译为"共振现象"）

又如：

activation 活化作用

distribution 分布状况

observation 观察结果

participation 参与管理

preparation 准备工作

resolution 解决方案

因此，熟悉介词与动词或形容词的习惯搭配是增词并正确理解词义的一种重要手段。

3. 分译

介词短语作定语时，往往是定语从句的一种简略形式。介词短语作状语时，有时是状语从句的简略形式。有些介词短语还是并列句的简略形式。因此汉译时，有的可以拆句分译。

（1）译成并列分句。如

Having been vaporized, the molecules of petroleum rise inside a fractionating tower, with the highest boiling point molecules liquefied first. 蒸发后，石油分子在分馏塔内上升，沸点最高的分子先液化。

（2）译成让步分句。如

An object may be hot without the motion in it being visible. 一个物体，即使其内部运动是看不见的，也可能是热的。

Man's warm blood makes it difficult for him to live long in the sea without some kind warmth. 人的血液是热的，如果得不到一定的热量，人就难以长期在海水中生活。

（3）译成结果状语从句。如

The resistance of this metal is too high for us to use it as a conductor of electricity. 这种金属的电阻太高，所以我们不能用它做导电体。（"for … electricity" 译为结果状语从句）

（4）译成原因分句。如

With more active catalysts, the reaction is carried very completely. 由于使用活性较大的催化剂反应进行得很完全。

4. 不译

不译或省略翻译是在确切表达原文内容的前提下使译文简练，合乎汉语规范，绝不是

任意省略某些介词。

(1) 表示时间或地点的英语介词,译成汉语如出现在句首,大都不译。如

For many years, Carbolite has been reducing the amount of asbestos used. 几年来,卡里播特公司一直在减少石棉的用量。

Many water power stations have been built in the country. 我国已建成许多水电站。

(2) 有些介词如 for(为了)、from(从……)、to(对……)、on(在……时)等,可以不译。如

The barometer is a good instrument for measuring air pressure. 气压计是测量气压的好仪器。

The air was removed from between the two pipes. 两根管子之间的空气已经抽出。

Answers to questions 2 and questions 3 may be obtained in the laboratory. 问题 2 和问题 3 的答案可以在实验室里得到。

Most substances expand on heating and contract on cooling. 大多数物质热胀冷缩。

(3) 表示与主语有关的某一方面、范围或内容的介词有时不译,可把介词的宾语译成汉语主语。

Something has gone wrong with the engine. 这台发动机出了毛病。

Gold is similar in color to brass. 金子的颜色和黄铜相似。

(4) 不少 of 介词短语在句中作定语。其中 of(……的)往往不译。如

The change of electrical energy into mechanical energy is done in motors. 电能变为机械能是通过电动机实现的。(of 短语和 change 在逻辑上有主谓关系,可译成主谓结构。)

Some of the properties of cathode rays listed below. 现将阴极射线的一些特性开列如下。(第一个 of 短语和 some 在逻辑上有部分关系, of 不译出。)

5. 反译

在不少情况下,有的介词短语如不从反面着笔,译文就不通,这时必须反译。如

(1) beyond, past, against 等表示超过某限度的能力或反对时,其短语有时用反译法。如

It is past repair. 这东西无法修补了。

There are some arguments against the possibility of life on this planet. 有些论据不同意这行星上可能有生物。

Radio telescopes have been able to probe space beyond the range of ordinary optical telescopes. 射电望远镜已能探测普通光学望远镜达不到的宇宙空间。

(2) off, from 等表示地点、距离时,有时用反译法。如

The boat sank off the coast. 这只船在离海岸不远处沉没了。

(3) but, except, besides 等表示除去、除外时,有时用反译法。如

Copper is the best conductor but silver. 铜是仅次于银的最优导体。

The molecular formula, C_6H_{14}, does not show anything except the total number of carbon and hydrogen atoms. 分子式 C_6H_{14} 只用来表示碳原子和氢原子的总数。

(4) from, in 等介词短语作补足语时,有时用反译法。如

An iron case will keep the Earth's magnetic field away from the compass. 铁箱能使地球磁场影响不了指南针。

The signal was shown about the machine being order. 信号表明机器没有毛病。

"一个词脱离上下文是不能翻译的（索伯列夫）"，没有上下文就没有词义。介词的翻译需根据上下文和词的搭配灵活处理，切忌作对号入座的机械翻译。

阅读技巧1：眼睛是如何阅读的

眼睛阅读实际上是指眼睛做一些较小的、有规律的"跳跃"。这些跳跃通常以比一个单词稍多一点的跨度使眼睛从一个凝视点跳到另一凝视点。由此看来，眼睛并不是光滑地在书面上扫描，而是以较小的弹跳从左向右移动，在继续移动和重复这些过程之前，先短暂地停顿一会儿，以吸收一个或两个单词。

阅读时的基本进程图

在进行阅读时，眼睛不断地重复移动、暂停、移动、暂停的动作。它只有在停顿期间才能吸收信息，这些暂停用去了大部分时间。每次暂停可能会延续四分之一秒到一秒半，因此，可以通过缩短每次停顿时间的方式提高阅读速度。

阅读技巧2：大脑后部的眼睛

由视网膜光接收器解码而得到的图像非常复杂，它沿着光神经被传输到大脑的可见区域——枕骨叶片。枕骨叶片并不紧靠眼睛后部，而是位于大脑的后部。因此，当听到一个人具有非常敏锐的观察力是因为在他的大脑后部有眼睛时，就没有什么值得大惊小怪的。控制阅读，并引导眼睛在书面中捕获感兴趣信息的正是位于大脑后部的枕骨叶片。这一认识为下面将要阐述的快速阅读方法奠定了基础。了解了有关眼睛的这些功能后，人们就会清楚地认识到：传统的阅读习惯和阅读速度是错误训练和误用的结果，因而，如果人们能更好地了解眼睛的功能并对其进行正确的训练，其功能将会得到有效的改善。

Reading Material

Determination of Hydrogen Peroxide (0.1%~6%) by Iodometric Titration

Principle

H_2O_2 oxidizes iodide to iodine in the presence of acid and molybdate catalyst. The iodine formed is titrated with thiosulfate solution, incorporating a starch indicator.

$$H_2O_2 + 2KI + H_2SO_4 \longrightarrow I_2 + K_2SO_4 + 2H_2O$$

$$I_2 + 2Na_2S_2O_3 \longrightarrow Na_2S_4O_6 + 2NaI$$

Scope of Application

This method is somewhat less accurate than the permanganate titration, but is less susceptible to interferences by organics, and is more suitable for measuring mg/L levels of H_2O_2.

Interferences

Other oxidizing agents will also produce iodine, whereas reducing agents (and unsaturated organics) will react with the liberated iodine. The contribution from other oxidizing agents can be determined by omitting the acid and molybdate catalyst.

Safety Precautions

Concentrated sulfuric acid is a corrosive, hazardous material and should be handled and disposed of in accordance with precaution. Neoprene gloves and monogoggles are recommended, as is working under a vacuum hood.

Sample bottles containing H_2O_2 should not be stoppered, but rather vented or covered loosely with aluminum foil or paraffin film.

Reagents

- Potassium iodide solution (1% W/V): Dissolve 1.0 grams KI into 100 ml demineralized water. Store capped in cool place away from light. Yellow-orange tinted KI solution indicates some air oxidation to iodine, which can be removed by adding 1~2 drops of dilute sodium thiosulfate solution.
- Ammonium molybdate solution: Dissolve 9 grams ammonium molybdate in 10 ml 6 mol·L^{-1} NH_4OH. Add 24 grams NH_4NO_3 and dilute to 100 ml.
- Sulfuric acid solution: Carefully add one part H_2SO_4-98% to four parts demineralized water.
- Starch indicator.
- Sodium thiosulfate solution (0.1 mol·L^{-1}).

Apparatus

- Analytical balance (±0.1 mg)
- Small weighing bottle (< 5 ml)
- 250 ml Erlenmeyer flask
- 50 ml buret (Class A)
- Medicine dropper

Procedure

(1) Weigh to the nearest 0.1 mg an amount of H_2O_2 equivalent to a titer of 30 ml (0.06 grams of H_2O_2) using a 5 ml beaker and medicine dropper. Transfer sample to Erlenmeyer flask.

(2) Add to Erlenmeyer flask 50 ml of demineralized water, 10 ml of sulfuric acid solution, 10~15 ml of potassium iodide solution, and two drops ammonium molybdate solution.

(3) Titrate with 0.1 mol·L^{-1} sodium thiosulfate to faint yellow or straw color. Swirl or stir gently during titration to minimize iodine loss.

(4) Add about 2 ml starch indicator, and continue titration until the blue color just disappears.

(5) Repeat steps (2)～(4) on a blank sample of water (omitting the H_2O_2).

Calculation

$$\text{Weight \% } H_2O_2 = \frac{(A-B) \times (\text{Normality of } Na_2S_2O_3) \times 1.7}{\text{Sample weight in grams}}$$

where A——Volume of $Na_2S_2O_3$ for sample, ml;

B——Volume of $Na_2S_2O_3$ for blank, ml。

Selected from "Haugland, R.P., Handbook of Fluorescent Probes and Research Chemicals; Molecular Probes Inc.; Eugene, OR, 1985."

PART 1 CHEMICAL ANALYSIS

Unit 6

Text: Complexometric Titration of Zn (II) with EDTA

Unknown Solution

Submit a clean 250 ml volumetric flask to the instructor so that an unknown zinc solution may be issued. Your name, section number, and your locker number should be written legibly on this flask. Note that the flask must be turned in at least 1 lab period before you plan to do the experiment, so that the Teaching assistants will have time to prepare the unknown.

Preparation of Solutions

1. EDTA, 0.01 mol · L^{-1}

Prepare at least one day ahead of time to make sure that the solute is dissolved. Dissolve about 3.8 g of the dihydrate of the disodium salt ($Na_2H_2Y · 2H_2O$) and 0.1 g $MgCl_2$ in 1 liter of water. Store in a plastic bottle. A small amount of sodium hydroxide may be added if there is any difficulty in dissolving the EDTA. Try not to exceed 3.8 g of $Na_2H_2Y · 2H_2O$ because much more than this dissolves only with difficulty. The EDTA solution should be filtered using suction filtration.

2. Buffer, pH 10

Each titration will require the addition of pH 10 ammonia buffer. The stock buffer solution has been prepared for you, and the appropriate quantity is dispensed directly into your titration flask from the "Repipet" repetitive dispenser located in hood. The buffer should be added immediately before you titrate a sample.

3. Calcium Standard Solution

A Ca^{2+} solution is prepared as the primary standard. Obtain approximately 0.7 g of predried analytical-reagent-grade $CaCO_3$. Accurately weigh a 0.25 g sample by difference into a 100 ml beaker. Add about 25 ml water and then add dilute HCl dropwise until the sample dissolves, then add 2 drops more. Mild heating will speed dissolution if necessary. Transfer quantitatively to a 250 ml volumetric flask (rinse well with water) and carefully dilute to the mark. (eye dropper!) Mix thoroughly. Because this Ca^{2+} standard solution is used to standardize the EDTA titrant, it must be prepared very carefully so that you know its exact molarity, and therefore the exact (to ±0.1 mg) mass of Ca_2CO_3 weighed out.

Standardization of EDTA Solution

(1) Pipet 25 ml aliquots of the standard Ca^{2+} solution into three or four 250 ml flasks. Remember that each aliquot contains one-tenth of the $CaCO_3$ weighed out to prepare the standard solution.

(2) Take each sample to completion before starting next sample. Add 7~8 ml of pH 10 buffer, 15 ml of water, and 3 drops of Eriochrome Black T and titrate immediately with EDTA until the light red solution turns a light sky blue.

Titrations must be performed swiftly (but carefully) because ammonia will evaporate and thus the pH of the solution will change. In general, the faster the titrations are performed, the better the results will be (as long as the end point is not overshot due to the speed). It is advantageous to perform a trial titration to locate the approximate endpoint and to observe the color change. In succeeding titrations, titrate very rapidly to within about 1 or 2 ml of the endpoint, then titrate very carefully (dropwise) to the end point.

(3) Calculate the molarity of the EDTA from the volume of EDTA used in the titration of each aliquot.

$$M_{EDTA} = \frac{\left[\dfrac{g\ CaCO_3}{10}\right]}{(\text{molar mass } CaCO_3) \times (V, \text{liters EDTA})}$$

The values (M_{EDTA} and titration volumes) should all agree very closely. If not, titrate additional aliquots until agreement is reached, and any spurious values can be rejected with confidence.

Determination of Zinc

(1) Dilute your unknown sample in the 250 ml volumetric flask to the mark with water. Mix thoroughly.

(2) Pipet 25.00 ml aliquots into 250 ml flasks. Add 15 ml of water, 9~10 ml of pH 10 buffer, and 2 drops of Eriochrome Black T immediately prior to titrating a sample.

(3) Titrate with standardized EDTA until the red solution turns blue.

(4) Calculate the number of milligrams of zinc in the total sample. Remember that each aliquot represents one-tenth of the total sample volume.

$$\text{mg Zn} = [\text{mmol Zn}] \times \left(\frac{1\ \text{mol}}{1000\ \text{mmol}}\right) \times \left(\text{molar mass Zn}, \frac{g}{\text{mol}}\right) \times (1000\ \text{mg/g}) \times 10$$
$$= M_{EDTA}(V, \text{ml EDTA}) \times (\text{molar mass Zn}) \times 10$$

Hazardous Waste Disposal

(1) Empty all titrated Ca and Zn solutions into the proper Hazardous Waste Bottle for this experiment, which is located under the central hood in the lab.

(2) When completely done with the experiment, mix any remaining EDTA titrant and any remaining Ca stock solution together in a large beaker. Pour down the drain with copious amounts of cold tap water. The two solutions are slightly basic and slightly acidic, respectively; when mixed, they will be near neutral. There are also no toxic chemicals present, so disposal directly down the drain is allowable and safe.

Notes
1. Eriochrome Black T Indicator

The color change of Eriochrome black T at the end point is rather subtle. It is not an abrupt change from bright red to a dark blue; but rather it is from a light red (or pink) to a pale blue. At least one trial titration is recommended. (You can always discard a "bad" value when you know there is a definite reason for its being bad.)

If you have trouble distinguishing the end point, a "before" and an "after" flask are recommended. Prepare two 250 ml flasks exactly the same as for the samples, except use distilled water instead of a sample; add additional distilled water to approximately equal the volume of EDTA titrant that would be titrated into the flask for the sample. Add the indicator. To one flask (the "after" the endpoint flask) add a small amount of EDTA to just past the color change at the endpoint. Stopper the flasks and keep them nearby for comparison of the colors. Titrate against a white background for better discrimination of colors. (Extra flasks can be checked out from the stockroom.)

Sometimes the Eriochrome Black T solution goes bad because of air oxidation. If the end points seem very indistinct to you, try a fresh bottle of indicator. Alternatively, try adding a small amount of solid Eriochrome Black T mixture (1 g indicator ground with 100 g NaCl); a small amount on the end of a spatula is sufficient.

2. pH 10 Ammonia Buffer

Dissolve 64.0 g of ammonium chloride in 600 mL of concentrated ammonia. Slowly and carefully add 400 ml deionized water. This should be sufficient for over 120 titrations.

Selected from "Brown, P. R. Analytical Chemistry, 1990, Vol. 62, pp. 995~1008.*"*

New Words

volumetric flask （容）量瓶
zinc   *n.* 锌
legibly   *adv.* 易读地，明了地
dihydrate   *n.* ［化］二水合物
disodium salt   二钠盐
filter   *vt.* 过滤，渗透，用过滤法除去
suction   *n.* 抽气机，抽水泵，吸引
buffer solution   缓冲溶液
ammonia   *n.* ［化］氨，氨水
calcium   *n.* ［化］钙（元素符号 Ca）
primary standard   基准物，原标准器
beaker   *n.* 大口杯，有倾口的烧杯
dropwise   *adv.* 逐滴地，一滴一滴地
quantitatively   *adv.* 数量上
Eriochrome Black T   铬黑 T
swiftly   *adv.* 很快地，即刻
evaporate   *v.* （使）蒸发，消失
milligram   *n.* 毫克

hazardous　*adj*. 危险的，冒险的，碰运气的
drain　*n*. 排水沟，排水
tap water　自来水，非蒸馏水
abrupt change　急剧（突然）变化
distilled water　*n*. 蒸馏水
deionize　*v*. ［物］除去离子，消电离

Notes

1. A small amount of sodium hydroxide may be added if there is any difficulty in dissolving the EDTA.

    参考译文：如果 EDTA 较难溶解，就加入少量的氢氧化钠。

2. Because this Ca^{2+} standard solution is used to standardize the EDTA titrant, it must be prepared very carefully so you know its exact molarity, and therefore the exact (to ± 0.1 mg) mass of Ca_2CO_3 weighed out.

    参考译文：因为 Ca^{2+} 标准溶液用来标定 EDTA 滴定剂，所以我们必须认真地配制 Ca^{2+} 标准溶液才能得知 EDTA 准确的摩尔浓度，因而要准确称量 Ca_2CO_3 的质量（精确到 ± 0.1 mg）。

3. Titrations must be performed swiftly (but carefully) because ammonia will evaporate and thus the pH of the solution will change. In general, the faster the titrations are performed the better the results will be (as long as the end point is not overshot due to the speed). It is advantageous to perform a trial titration to locate the approximate endpoint and to observe the color change.

    参考译文：我们必须迅速而仔细地进行滴定。这是由于氨水会蒸发，溶液的 pH 因此会发生变化。一般来说，滴定速度进行得越快，滴定结果就越好（只要滴定终点不会因为速度的缘故而滴过）。这样有利于近终点滴定操作并且便于观察颜色的变化。

4. Sometimes the Eriochrome Black T solution goes bad because of air oxidation. If the endpoints seem very indistinct to you, try a fresh bottle of indicator. Alternatively, try adding a small amount of solid Eriochrome Black T mixture (1 g indicator ground with 100 g NaCl); a small amount on the end of a spatula is sufficient.

    参考译文：有时铬黑 T 溶液由于被空气氧化而变质。如果在你看来滴定终点很模糊，就换一瓶新配制的指示剂。此外你也可以加入少量的固体铬黑 T 混合物（1 g 铬黑 T 碎末混有 100 g NaCl），只要刮刀刀尖端上的一点点混合物就足够了。

Exercises

1. Put the following into Chinese.

disodium salt	sodium hydroxide
mix thoroughly	calcium
primary standard	Eriochrome Black T
ammonium chloride	deionized water

2. Put the following into English.

容量瓶	锌

缓冲溶液   氨水
自来水   蒸馏水

3. Questions.

(1) How long do we have to have the volumetric flask in complexmetric titration of Zn(Ⅱ) with EDTA?

(2) Why can't we try to exceed 3.8 g of $Na_2H_2Y \cdot 2H_2O$ when we prepare the 0.01mg EDTA?

(3) How many steps are there in the standardization of EDTA solution?

(4) What if you have trouble distinguishing the end point with Eriochrome Black T indicator?

(5) How can you deal with the situation in which the Eriochrome Black T solution goes bad because of air oxidation?

科 技 英 语 基 本 技 巧
翻译技巧：英语数字的翻译

英语中有些数词在汉译时可以等值翻译。但是，也有不少数词在汉译中不能等值翻译，或者完全不译出来。这种翻译处理方法是为了使汉译句子能符合汉语的表达习惯。以下分别举例说明。

一、量的译法

1. 用阿拉伯数字表示的数量

这在科技英语中是最常见的，其译法不外三种。

(1) 照抄　数字不太大（例如在五、六位数以下）的，包括温度、压力、年份、产量、耗量、金额等，一般照抄。例如

at 250℃   在 250℃
by 1980   到 1980 年
15380 tons   15380 吨

(2) 换算　较大的数字可以加以换算，利用"万""亿"等计数单位加以表示。因为欧美各国沿用千位制，而中国习惯于使用万位制。所以对较大的数字要适当地进行换算，这样既可以将数字简化，又符合中国的通用叫法。例如

5000000 tons   500 万吨
23000000 tons   2300 万吨
78546000 tons   7854.6 万吨

(3) 译成中文　较大的整数（万的整数倍，而且只有一位有效数字），在单独出现（其后跟有单位名称）时，可以全部译成中文。例如 $5000000 可以翻译成"五百万美元"或

"500万美元",但是不能翻译成"5百万美元"。(汉语中不用"百万""千万"做计数单位)。如果和其他数字并列。而别的数字不能这样翻译时,一般应一律照抄。

2. 用文字表示的数量

(1) 确定的数量可以根据具体情况译成数码和汉语。

    four hundred and fifty   450 或四百五
    twenty-three thousand and six hundred   23600(两万三千六百)
    three hundred thousand   三十万
    five quadrillion 5×10^{15}
    eight quadrillion 8×10^{18}
    five three thousand four hundred and eighty (dollars) only 伍万叁仟肆佰捌拾美元整(金额用大写表示)。

在此值得注意的是,英国和德国都采用是英国制计数法,而法国和美国等国家则采用大陆制计数法。前者是百万进位制,后者是千进位制。因此等数词在两种计数法里有不同的含义。遇到这样的大数时,需先搞清楚原文是何种计数法,才能正确地加以翻译。

(2) 大约的数量 这有下列几组表示法。

  ① tens of   数十(几十)
    decades of   数十年(几十年)
    dozens of   几十(几打)
    scores of   许多
    hundreds of   几百
    tens and thousands of   数万(几万)
    a hundred and one   无数的,许多的
    millions of   千千万万,千百万,成千上万,亿万
    ten to one   十之八九,十有八九(插入语)
    the seventies   七十年代
    the early 1980'   二十世纪八十年代
    up to 2.5 vol%   ≤2.5%(体积分数)
    down to 480kg/t   ≥480 公斤/吨

  ② 数字后加 "odd", "and odd" 或 "or odd"
    twenty [and] odd   二十几个,二十有余
    four hundred [or] odd   四百多

  ③ 数字前加 "more than" 或 "over", "above"
    more than thirty   三十多,三十以上,高于三十
    over 30   三十多,三十以上,高于三十
    above 30   三十多,三十以上,高于三十

  ④ 数字前加 "less than" 或 "below"、"under"
    less than 40   四十以下
    under 40   不到四十
    below 40   低于四十

⑤ 数字前加 some 或 about
  some 50 pounds   大约五十磅
  about 50 lbs   五十磅左右
⑥ 数字后加 or so、or more、or less、more or less
  60 grams or so   60 克左右（上下）
  60 grams or more   60 克以上
  60 grams or less   60 克以下
  2 hours more or less 两小时左右
⑦ 利用介词 to 和 between 表示数量范围
  70 to 80   七、八十
  from 70 to 80   从 70 到 80，70～80
  between 70 and 80   介于 70 到 80 之间，70 到 80
⑧ 用 upwards of, close to, approximate to, of the order of, order of magnitude, with a factor of ten 等短语
  upwards of 80 years   八十多年
  close to a hundred   近一百年
  approximate to a thousand 近千，约一千
  of the order of 5percent 大约 5%
  three orders of magnitude 三个数量级
  with a factor of ten   在十个数量级的范围之内

二、增加量的译法

(1) The output went up 56000 tons. 产量增长了 56000 吨

(2) The production has increased by 36%. 产量已经增加 36%。

(3) In 1975, the output value of Shanghai's heavy industry multiplied 18 times as against 1949. 1975 年与 1949 年相比，上海重工业产值增长了 17 倍。

(4) The liquor was diluted with water to five times its original volume. 溶液用水稀释到其原来体积的五倍

(5) Chromium masks lasts 10 to 100 times longer than the emulsion masks. 铬掩膜的使用寿命为乳胶掩膜的 10～100 倍。

三、减少量的译法

英语里有"减少了多少倍"和"成多少倍地减少"的说法，但汉语里不能这样说，只能译成"减少了（或减少到）几分之几。"大体上，"成 n 倍地减少"可译为"减少到 $\frac{1}{n}$（当 n 为整数时）"或"减少了 $\frac{n-1}{n}$（$n \leqslant 10$）"。而"减少了 n 倍"则应该译为"减少到 $\frac{1}{n+1}$""减少了 $\frac{n}{n+1}$"（$n \leqslant 9$）。

(1) 净减量，所减的数字可以照译。
  decrease by 10   减去 10
  reduce by 10%   减少 10%
  fall by 60%   下降 60%
  35% less   少 35%

(2) 成 n 倍地减少，即减少前的数量是减少后的数量的 n 倍（减少成 $1/n$），译时需要换成分数，译为"减少到 $1/n$"或"减少了 $\frac{n-1}{n}$"。

  decrease 2 times   减少到 1/2，减少了 1/2
  shorten 30 times   缩减到 1/30
  a six-fold difference   相差五倍
  10 times as light as   比……轻十分之九

(3) 减少了一半。

  decrease one-half   减去一半
  reducing by one-half   减小一半
  shorten two times   缩短一半
  halving   把……减小一半
  one-half less   少一半，小一半
  be less than half   少一半还多

(4) 减少了 n 倍，即减少前的数量比减少后的数量大 n 倍。译时需要换算成分数，译为"减少了 $\frac{1}{n+1}$"。

  4 times less (shorter, lighter)   少（短、轻）4/5
  twice less   少 2/3，小 2/3
  twice thinner   厚度减薄了 2/3

(5) decrease to 50   减小到 50
  reduce to 60%   减少到 60%
  cut to 70%   降低到 70%

四、序数词的译法

(1) 表示时间和空间的顺序，其译法举例如下。

  nineteen nineties   20 世纪 90 年代
  the first time   第一次
  the first plants   头一批工厂

(2) 表示分数，其译法举例如下。

  two thirds   2/3
  one hundredth   百分之一
  one-second power   1/2 次幂
  a few millionths of a second   百万分之几秒

阅读技巧 1：较差阅读者

  图 6-1 显示了那些较差阅读者的眼睛运动情况。这类读者阅读时停顿次数比理解程度好的读者的停顿多。这种停顿也就是通常所说的"凝视"。造成这些额外停顿的原因，是那些较慢的读者常常重复读单词，有时往回跳两到三处，其目的是能正确理解单词的意思。这些回跳（几乎是习惯地回到刚看过的单词上）习惯和倒退阅读（有意识地回到那些被认为漏掉或误解的单词上），使得那些较差阅读者过多使用了凝视。

  研究表明：在 80% 的情况下，当不进行回跳或倒退时，读者发现自己实际上已经吸收了那些信息（是在他们开始阅读下一个词组时开始吸收的），快速阅读者很少纵容这些会大大

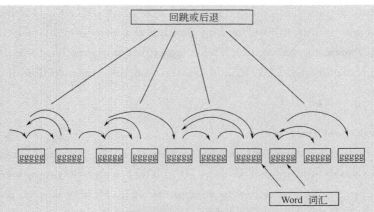

图 6-1 较差读者的眼睛运动示意

降低阅读速度的不必要重复。如果每次回跳或倒退浪费 1 min 20 s，以一本 300 页的书计算，读者就将有 $6\frac{2}{3}$ h 被额外浪费在回跳或倒退阅读上，这种重复阅读几乎没有任何意义。

阅读技巧 2：快速阅读者

图 6-2 显示了那些快速者在没有回跳和倒退的情况下，能以较大幅度的跳跃进行阅读。他们每次吸收的不是一个单词，而是三四个甚至五个单词。

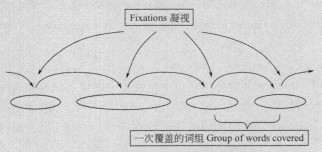

图 6-2 优秀读者的眼睛运动示意

下面我们将快速阅读者与较慢阅读者比较一下。我们假定每次凝视都需要半秒钟。以阅读 12 个单词的行计，较慢的阅读者每次凝视一个单词，并且回跳或倒退两次，那么他将需要 14 个 0.5 s，即 7 s；而快速阅读者每次凝视三四个单词，并且没有回跳或倒退，则只需要两秒钟。快速阅读者对使用眼睛进行阅读的技巧稍作调整之后，其阅读量和阅读速度就将是较慢阅读者的 3.5 倍。

Reading Material

Determination of Water Hardness by Complexometric Titration

Hard Water

Hard water is due to metal ions (minerals) that are dissolved in the ground water. These minerals include Ca^{2+}, Mg^{2+}, Fe^{3+}, SO_4^{2-}, and HCO_3^-. Our hard water in the southern Indiana area is due to rain moving through the vast amount of limestone, $CaCO_3$ that occurs in our area to the aquifer. This is why we measure hardness in terms of $CaCO_3$. The concentration of the Ca^{2+} ions is greater than the concentration of any other metal ion in our water.

Why be Concerned about Hard Water

The determination of water hardness is a useful test that provides a measure of quality of water for households and industrial uses. Originally, water hardness was defined as the measure of the capacity of the water to precipitate soap. Hard water is not a health hazard. People regularly take calcium supplements. Drinking hard water contributes a small amount of calcium and magnesium toward the total human dietary needs of calcium and magnesium. The National Academy of Science states that consuming extremely hard water could be a major contributor of calcium and magnesium to the diet.

Hard water does cause soap scum, clog pipes and clog boilers. Soap scum is formed when the calcium ion binds with the soap. This causes an insoluble compound that precipitates to form the scum you see. Soap actually softens hard water by removing the Ca^{2+} ions from the water.

When hard water is heated, $CaCO_3$ precipitates out, which then clogs pipes and industrial boilers. This leads to malfunction or damage and is expensive to remove.

Water Softeners

If you have hard water you may use a water softener to remove the hardness. Salt is mixed with water. The Na^+ ion from the salt replaces the Ca^{2+} ion, but this causes the water to be too salty for drinking. Water that has been softened should be used only for laundry and bathing.

Types of Hardness

There are two types of water hardness, temporary and permanent.

Temporary Hardness is due to the bicarbonate ion, HCO_3^-, being present in the water. This type of hardness can be removed by boiling the water to expel the CO_2, as indicated by the following equation:

$$HCO_3^- \rightleftharpoons OH^- + CO_2(g)$$

Bicarbonate hardness is classified as temporary hardness.

Permanent hardness is due to the presence of the ions Ca^{2+}, Mg^{2+}, Fe^{3+} and SO_4^{2-}. This type of hardness can't be eliminated by boiling. The water with this type of hardness is said to be permanently hard.

Complexometric Titration

Permanent hardness is usually determined by titrating it with a standard solution of ethylenediamminetetraacetic acid, EDTA. The EDTA is a complexing, or chelating agent used to capture the metal ions. This causes the water to become softened, but the metal ions are not removed from the water. EDTA simply binds the metal ions to it very tightly.

EDTA

EDTA is a versatile chelating agent. A chelating agent is a substance whose molecules can form several bonds to a single metal ion. Chelating agents are multi-dentate ligands. A ligand is a substance that binds with metal ions to form a complex ion. Multi-dentate ligands are many clawed, holding onto the metal ion to from a very stable complex. EDTA can form four or six bonds with a metal ion.

$$\begin{array}{c} HOOC-CH_2 \diagdown \qquad\qquad \diagup CH_2-COOH \\ :N-CH_2-CH_2-N: \\ HOOC-CH_2 \diagup \qquad\qquad \diagdown CH_2-COOH \end{array}$$

Structure of EDTA

It is frequently used in soaps and detergents because it forms complexes with calcium and magnesium ions. These ions, which are in hard water are bound to the EDTA and cannot interfere with the cleaning action of the soap or detergent.

EDTA is also used in foods. Certain enzymes are responsible for food spoilage. EDTA is used to remove metal ions from these enzymes. It is used to promote color retention in dried bananas, beans, chick peas, canned clams, pecan pie filling, frozen potatoes and canned shrimp. It is used to improve flavor retention in canned carbonated beverages, beer, salad dressings, mayonnaise, margarine, and sauces. It inhibits rancidity in salad dressings, mayonnaise, sauces and salad spreads.

Total Permanent Hardness

In this lab you will be asked to determine the total permanent hardness. EDTA grabs all the metal ions in the water, not just the Ca^{2+} ions. This gives us a value that is not truly the concentration of Ca^{2+} ions. This causes an experimental error of about 1%, which is acceptable due to the "fuzzy" endpoints in this type of titration.

EDTA End Point Color Change

Erio T indicator or Eriochrome Black T indicator is used in this titration. When it is chelated or acidified, it produces a wine red solution. When it is not chelated and under basic conditions it is blue.

Selected from "Brown, P. R. Analytical Chemistry, 1990, Vol. 62, pp. 995~1008."

PART 2　INSTRUMENT ANALYSIS

Unit 7

Text: Determine Permanganate by Spectrophotometric Analysis

Introduction

In this experiment, a series of solutions, each of which has different but known concentrations, will be prepared. The "absorption spectrum" of the permanganate will be using one of these solutions, i.e., a plot of light absorbed by the species as a function of wavelength (λ). The wavelength showing the strongest absorption (λ_{max}) will be determined. Measuring λ_{max} for each permanganate solutions enables the calibration of absorbance of light to the concentration of the permanganate species. You will use this calibration to determine the concentration of a permanganate solution.

Spectroscopy

In this experiment, two different oxidation states of the element manganese (Mn) will be encountered. One of the states is the Mn^{2+} species; this ion is stable in an acid electrolyte and colorless (does not absorb light in the visible region of the spectrum). Treatment of the Mn^{2+} solution with a strong oxidant such as potassium periodate (KIO_4) converts the Mn^{2+} to a purple permanganate species.

$$2Mn^{2+} + 5IO_4^- + 3H_2O \longrightarrow 2MnO_4^- + 5IO_3^- + 6H^+$$

The characteristic purple color of the MnO_4^- results from absorption of the green-yellow components of the visible spectrum.

Operating the Mini-Spectrophotometer

- The control knob allows you to select $0\% \ T$ or one of three wavelengths (colors) of light. The $0\% \ T$ is used only during the calibration step. With this selection, the light is blocked completely so that none reaches the photocell, irrespective of what sample is in the cuvette.

- Non-absorbing solutions have $T=1$ ($100\% \ T$) and $A=0$ while complete absorption of light gives $T=0$ ($0\% \ T$) and $A=\infty$. The absorbance of species X in solution is proportional to its concentration and the path length. The relation is expressed in the form of the Beer-Lambert law.

$$A_x = \varepsilon_x b c_x$$

• Always handle the cuvettes by the top and bottom edges. The cuvettes have two different pairs of sides, one pair of sides are clear for light transmittance and the other pair of sides are ridged for easier manipulation. Never touch the clear sides, except to clean them with lens paper.

• Cuvettes are filled three-quarters and the sides are wiped down by using Kimwipes. When placing the cuvette in the colorimeter, the clear sides must line up with the light source (denoted by the white line). The total materials and equipment is given in Table 7-1.

Table 7-1 Materials/Equipment

top-loader	colorimeter
600 ml beaker	wire gauze
Bunsen burner	distilled water
50 ml burette	15 mol·L^{-1} H_3PO_4 solution
retort stand	KIO_4 solution
six 100 ml volumetric flasks	Mn^{2+} solution
one plastic cuvette	

Procedure
Preparation of Standard and Unknown Solutions

You will prepare four standard Mn^{2+} solutions as well as two, identical Mn^{2+} solutions of unknown concentration. Accurately measure the required volumes using a 50 ml burette.

(1) Use a burette to measure out 5.00 ml, 15.00 ml, 25.00 ml and 40.00 ml volumes of the standard Mn^{2+} solution into each of four Pyrex 100 ml volumetric flasks (label them using masking tape).

(2) Use a burette to measure out two 25.00 ml portions of the unknown Mn^{2+} solutions into each of two Pyrex 100 ml volumetric flasks (label them using masking tape).

(3) Handling with care, add 5 ml of concentrated phosphoric acid (15 mol·L^{-1} H_3PO_4) to each flask from the dispenser.

(4) Use the top-loading (top-loader) balance to add approximately 0.4 g of potassium periodate (KIO_4) to each flask.

(5) Dilute the solution to approximately 80 ml (about 2/3 full) using distilled water.

(6) Mix the contents of each flask by swirling. Working with the flasks one at a time (starting with the most concentrated solution)

(7) Heat the flask in a boiling water bath using a 600 ml beaker half-filled with tap water and 3 or 4 boiling chips heated over a Bunsen burner. The purple color of permanganate will develop as the oxidation proceeds.

(8) After 10 minutes of heating, remove the flask from the hot water bath. Allow the flask to cool slowly at room temperature to about 50℃ then, using cold water, bring the flask and its contents to near room temperature.

(9) Use distilled water to make the solution up to the mark, i.e., add the water until the level of the meniscus is just up to the mark (not over the mark). Mix well by closing the

flask with a rubber stopper and shaking and inverting the flask for several minutes. You now have exactly 100 ml±0.01 ml of solution.

Clean Up

Upon completion of the lab, do not forget to clean up.
- Dilute the excess reagents well and dump them down the sinks.
- Wash the glass containers with the soapy water provided in the dishpans beside the sinks. Rinse the containers with tap water then with distilled water.
- Return all the equipment to its proper place.
- Wipe your work area clean and dry.

Selected from "E. B. Sandell, Colorimetric Determination of Traces of permanganate, 3rd Ed. Interscience Publishers, Inc., New York, 1959."

New Words

absorption spectrum 吸收光谱，吸收频谱
permanganate *n*. [化] 高锰酸盐
wavelength *n*. 波长
absorption *n*. 吸收
spectroscopy *n*. [物] 光谱学，波谱学，分光镜使用
manganese *n*. [化] 锰（元素符号 Mn）
stable *adj*. 稳定的
electrolyte *n*. 电解，电解液
periodate *n*. [化] 高碘酸盐
visible spectrum 可见光谱
spectrophotometer *n*. 分光光度计
photocell *n*. 光电池
cuvette *n*. 比色皿，透明小容器，试管
proportional *adj*. 成比例的，相称的，均衡的
path length 路径长度
Beer-Lambert law 朗伯-比耳定律
manipulation *n*. 处理，操作，操纵，被操纵
lens *n*. 透镜，镜头
wire gauze 金属细网纱
phosphoric *adj*. [化] 磷的（尤指含五价磷的），含磷的
room temperature 室温，常温（约 20℃）
meniscus *n*. 新月，弯液面，凹凸透镜

Notes

1. In this experiment, a series of solutions, each of which has different but known concen-

trations, will be prepared.

  参考译文：在使用分光光度计来测定高锰酸根的实验中，我们要配制一系列不同浓度的标准溶液。

2. Treatment of the Mn^{2+} solution with a strong oxidant such as potassium periodate (KIO_4) converts the Mn^{2+} to a purple permanganate species.

  参考译文：采用像高碘酸钾（KIO_4）这样的强氧化剂来氧化 Mn^{2+}，可以使 Mn^{2+} 转化为紫色的高锰酸根。

3. Non-absorbing solutions have $T=1$ ($100\% \ T$) and $A=0$ while complete absorption of light gives $T=0$ ($0\% \ T$) and $A=\infty$. The absorbance of species X in solution is proportional to its concentration and the path length. The relation is expressed in the form of the Beer-Lambert law.

  参考译文：溶液不吸光时，透光率等于 1（$100\% \ T$），吸光度为零，而溶液完全吸光时，透光率等于零（$0\% \ T$），吸光度为无穷大。含某物质 X 溶液的吸光度与该溶液的浓度及路径长度成比例。这种关系可以用朗伯-比耳定律来表示。

4. Use distilled water to make the solution up to the mark, i. e., add the water until the level of the meniscus is just up to the mark (not over the mark). Mix well by closing the flask with a rubber stopper and shaking and inverting the flask for several minutes. You now have exactly 100 ml±0.01 ml of solution.

  参考译文：滴加蒸馏水使溶液达到标线，即滴加蒸馏水直至弯液面水平面刚刚达到标线（但不得超过标线），用橡胶塞子盖紧容量瓶，将溶液和蒸馏水充分混合均匀，将容量瓶倒置并振摇数次。你即可准确得到 100 ml± 0.01 ml 溶液。

Exercises

1. Put the following into Chinese.

manganese	absorption spectrum
strong oxidant	potassium periodate
spectrophotometer	Beer-Lambert Law
lens paper	

2. Put the following into English.

波长	电解液
可见光谱	磷酸
试管	

3. Questions.

  (1) How many oxidation states of the element manganese (Mn) will be uncounted in this experiment?

  (2) What are the materials used in this experiment?

  (3) Can you tell us how to prepare standard and unknown solutions in simple English?

  (4) How can you go through the procedure of cleaning up in this experiment?

  (5) Do you know anything about the Beer-Lambert law?

科 技 英 语 基 本 技 巧
翻译技巧：被动语态的译法

英语中被动语态的使用范围极为广泛，尤其是在科技英语中，被动语态几乎随处可见，凡是在不必、不愿说出或不知道主动者的情况下均可使用被动语态。在汉语中也有被动语态，通常通过"把"或"被"等词体现出来，但它的使用范围远远小于英语中被动语态的使用范围，因此英语中的被动语态在很多情况下都翻译成主动结构。对于英语原文的被动结构，我们一般采取下列方法。

1. 翻译成汉语的主动句

英语原文的被动结构翻译成汉语的主动结构又可以进一步分为几种不同的情况。

(1) 英语原文中的主语在译文中仍做主语，但改变谓语的结构（如将动词改成动宾结构）。在采用此方法时，我们往往在译文中使用了"得到""受到""加以""经过""用……来"等词来体现原文中的被动含义。例如：

Growing needs for cryolite by the aluminium and other industries are increasingly being met via synthetic feed-material. 炼铝工业和其他工业对冰晶石日益增长的需求，正通过采用合成原料而不断地得到满足。

Sometimes the communication would be seriously disturbed by solar spots. 通讯有时会受到日斑的严重干扰。

In other words, mineral substances, which are found on earth must be extracted by digging, boring holes, artificial explosions, or similar operations, which make them available to us. 换言之，矿物就是存在于地球上，但需经过挖掘、钻孔、人工爆破或类似作业才能获得的物质。

Nuclear power's danger to health, safety, and even life itself can be summed up in one word: radiation. 核能对健康、安全，甚至对生命本身构成的危险可以用一个词——辐射来概括。

(2) 将英语原文中的主语翻译为宾语，同时增补泛指性的词语（人们，大家等）作主语。例如

Air can be liquefied by us. 我们能将空气液化。

Hydrogen is known to be the lightest element. 人们知道氢是最轻的元素。

It is just the energy which the atom thus yields up that is held to account for the radiation. 人们认为，这种辐射正是由原子释放出来的能量造成的。

另外，下列结构也可以通过这一方式进行翻译。

It is reported that… 有人报告，据报道……

It is stated that… 有人说，据说……

It is generally considered that… 大家（一般人）认为……

It is well known that… 大家知道（众所周知）……

It will be said… 有人会说……
It was told that… 有人曾经说……

（3）将英语原文中做状语的介词短语翻译成译文的主语，在此情况下，英语原文中的主语一般被翻译成宾语。例如

In the present study the rate constant based on mass transport of hydrogen in the metal phase was found to be inversely proportional to the two-thirds power of the bulk oxygen content in the three metals.

本研究发现，基于氢在金属相中的物质传递的速度常数，是和三种金属中氧的体积含量的 2/3 次幂成正比的。

A right kind of fuel is needed for an atomic reactor. 原子反应堆需要一种合适的燃料。

（4）翻译成汉语的无主句。例如

Several approaches to the problem of ladle skull, slag or deoxidation-scum removal are being tried. 罐内结壳、熔渣或脱氧浮渣清除问题的几项解决方案正在实验之中。

Sugar can be dissolved in water. 糖可溶解于水中。

Increased rate of reaction, in the process of catalyzes, is called catalysis, or catalytic action. 有催化剂时反应速率加快叫做催化作用。

By this procedure, different honeys have been found to vary widely in the sensitivity of their inhibit to heat. 通过这种方法分析发现不同种类的蜂蜜的抗菌活动对热的敏感程度也极为不同。

另外，下列结构也可以通过这一手段翻译。

It is hoped that… 希望……
It is reported that… 据报道……
It is said that… 据说……
It is supposed that… 据推测……
It may be said without fear of exaggeration that… 可以毫不夸张地说……
It must be admitted that… 必须承认……
It must be pointed out that… 必须指出……
It will be seen from this that… 由此可见……

2. 译成汉语的被动语态

英语中的许多被动句可以翻译成汉语的被动句。常用"被""给""遭""挨""为……所""使""由……""受到""让""叫"等表示。例如

Up to now, sulphur dioxide has been regarded as one of the most serious of these pollutants. 到目前为止，二氧化硫一直被看作是这些污染物中最严重的一种。

A machine is assembled of its separate components. 机器是由一些单独的部件装配成的。

Pure oxygen must be given to the patients in certain circumstances.
在某些情况下必须给病人吸纯氧气。

Natural light or "white" light is actually made up of many colors. 自然光或者"白光"实际上是由许多种颜色组成的。

The behavior of a fluid flowing through a pipe is affected by a number of factors, including the viscosity of the fluid and the speed at which it is pumped. 流体在管道中流动的

情况，受到诸如流体黏度、泵输送速度等各种因素的影响。

　　The supply of oil can be shut off unexpectedly at any time, and in any case, the oil wells will all run dry in thirty years or so at the present rate of use. 石油的供应可能随时会被中断；不管怎样，以目前的这种消费速度，只需 30 年左右，所有的油井都会枯竭。

　　This extraction rate was confirmed in batch tank tests. 这一提取速度已为分批槽内实验所证实。

　　Iron is extracted from the ore by smelting in the blast furnace. 铁是通过高炉冶炼从矿石里提取出来的。

阅读技巧：快速阅读训练

　　1. 计时阅读（Timed Reading）

　　语言学家桑卡（Sonka）说："如果你想要具有快速阅读的能力，其秘诀就是在计时条件下进行练习。"快速阅读材料应偏易，生词量、语法结构难度适中，必须是没有读过的文章（unseen passage），利用教材中的课文进行快速阅读训练，一定不要预习，即定时间要合理，要求经常自测 WPM 记录，了解进度情况。

　　2. 成组视读（Phrase Reading）

　　放宽"视幅"，防止回视，强化记忆功能，以词组或"意群"（thought group）为单位进行阅读，如：The little boy Johnnie \ had been up \ with a packet of mints \ and said \ he wouldn't go out to play \ until the post had come.

　　3. 略读（Skimming）

　　语言学家尼古拉·史密斯说："如果你想读得快，必须狼吞虎咽，而不是细嚼慢咽"。略读要求跳过枝叶抓主干，省去语言中冗余现象，如（冠词、形容词、插入语等），集中注意力找出中心词（key words），从而抓住中心思想。如：Her dear old friend Jane has a dog. 这一句中，只要抓住 Jane has a dog 就行了。略读不仅可以提高阅读速度一倍，还可有助于提高理解能力。

　　4. 找主题句（Topic Sentence）

　　阅读理解练习中，不但要求找出相关细节，还要求做出预测（prediction）和总结出中心思想（main idea）。这些深层次理解能力，也可以在快速阅读技巧训练中加以培养。在专业英语中，其主题句通常为该段落的第一句话，它可以告诉该段文字的中心和大意，所以对每段的第一个句子要予以重视。如果主题句不在一段之首，则可能在该段之末。总之，对于一篇文章、一个章节，要注意其首尾两段。通常情况是：首段开宗明义；末段概括总结。找准主题句，有助于学生对文章中心进行把握。

　　5. 猜词悟意（Word-Guessing）

　　阅读理解对生词进行考察越来越普遍，对于文章中出现的生词，不查字典，通过上下文联系、逻辑推理、构词法以及一些非语言知识，可以推词义和段落。常用技巧有定义（definition）、对比（contrast）等。

　　（1）A **satellite** is an object, either natural or man-made, which travels in an orbit around another object in space.（卫星，定义）

　　（2）My parents went out and bought a new TV set. That afternoon an **antenna** was put on the roof.（天线，常识）

Reading Material

Spectrophotometry

Spectrophotometric analysis is the most widely used method of quantitative and qualitative analysis in the chemical and biological science; it is an accurate and very sensitive method, which can analyze quantities as small as micrograms. The method depends on the light absorbing properties of either the substance being analyzed or one of its derivatives. The basis of spectrophotometry is simple: the intensity of the light which is transmitted through a solution containing an absorbing substance (chromogen) is decreased by that fraction which is absorbed, and this fraction can be detected and measured photoelectrically. The objective of this experiment is to apply the principle and technique of spectrophotometry to the determination of proteins and nucleic acids.

Beer-Lambert Law

The Beer-Lambert law states that the amount of light absorbed is proportional to the number of molecules of absorbing substance in the light path; i.e. absorption is proportional both to the concentration of the chromogen in solution and to the length of the light path through the solution. This relationship can be expressed as follows.

$$-\lg \frac{I}{I_0} = Kcl$$

where I and I_0 are the intensity of transmitted light in the presence and in the absence of the chromogen, respectively; c is the concentration of the chromogen; l is the length of the light path through the solution; K is a constant, characteristic for each absorbing substance at a specific wavelength of light and in a specified solvent.

The ratio I/I_0 is called the light transmission and is usually measured by percent. The absorbance (A), or optical density (OD), is the quantity more frequently used, and is given by

$$A(OD) = -\lg \frac{I}{I_0}$$

On substitution of this definition into the Beer-Lamber law; $A(OD) = Kcl$.

In this form the Beer-Lambert law states that doubling of either the concentration of the absorbing substance or doubling of the depth of solution leads to a doubling of the OD. The quantitative measurement of the amount of light absorbed is accomplished with an instrument called a spectrophotometer. Spectrophotometers detect photoelectrically and compare electronically the difference in the amount of light transmitted through solutions containing differing concentrations of an absorbing substance. The difference is expressed on a dial of the instrument as percent transmission or optical density.

In the expression for the Beer-Lambert law, the proportionality constant K depends on the wavelength of light and the nature of the absorbing substance. This relationship is the one used in spectrophotometric work with precise instruments, such as the Beckman spectrophotometer (Model DB or DB-G). In this case, that portion of the spectrum used is so narrow that it can be considered monochromatic (a single wavelength). In additon, carefully made solution containers, called cells or cuvettes, with plane parallel sides are used so that the light path is the same through all portions of the cell and is accurately known. In such a case, it is possible to tabulate values of the constant K for various substances in various solvents and wavelengths. If the concentration is expressed in moles per liter and the light path in cm, then the proportionality constant K is called the molar extinction coefficient. That is, the absorbency of a $1 \text{ mol} \cdot L^{-1}$ solution of 1 cm path length.

Less precise instruments, such as the Spectronic 20 colorimeters do not pass light of a single wavelength (monochro-matic light) but beam a broader band of the spectrum upon the sample. In applying the Beer-Lambert law to measurements obtained with these instruments, the constant K in the original equation is replaced by K thus:

$$OD = Kcl$$

The value of K depends on the spectral band width as well as an other parameters, and hence is a constant only for a specific type of colorimeter.

Quantitative Measurements

If, as will be the case in this course, the cells which contain the solutions to be analyzed are of constant diameter and the wavelength is not varied, K and l may be joined together to give a working constant K' and

$$OD = K'c$$

Since OD is directly proportional to concentration a plot of OD versus concentration yields a straight line, the slope of which represents the constant K'.

The quantitative measurement of an absorbing substance is accomplished using such a plot, called a standard curve. The curve is constructed by preparing a series of standards of the substance to be analyzed in a graded series of known concentrations. The OD of each standard is then determined and a plot constructed of the OD on the abscissa. The OD of a solution of unknown concentration is then determined under identical conditions and the value of the concentration is simply read from the plot.

Qualitative Measurements

As stated above absorption may occur when light of the energy (wavelength) required to excite certain of the bonding electrons strikes the absorbing molecule. The intensity and position in the electromagnetic spectrum of the absorbed light is related in a complicated way to the electronic structure of the molecule and its constituent atoms and is a unique optical characteristic of the substance. A plot of the OD of a solution of an absorbing substance over a range of wavelengths is referred to as the absorption spectrum of the compound and is useful for the qualitative identification of unknown compounds. Since the concentration of the chromogen is constant for such a plot the curve describes the dependence of K upon wavelength. Values for K are largest in regions referred to as absorption peaks or absorption bands. The wavelength, or l (lambda), of maximum absorption of a particular compound is usually referred to as the I_{max}.

Selected from "E. B. Sandell, Colorimetric Determination of Traces of Metals, 3rd Ed. Interscience Publishers, Inc., New York, 1959."

Unit 8

Text: Infrared Spectroscopy

Description

This exercise is intended to familiarize you with the identification of functional groups in organic compounds using infrared spectra. Before you can use this technique, you need to have an introduction to infrared spectroscopy and to what an IR spectrum is.

Molecules are flexible, moving collections of atoms. The atoms in a molecule are constantly oscillating around average positions. Bond lengths and bond angles are continuously changing due to this vibration. A molecule absorbs infrared radiation when the vibration of the atoms in the molecule produces an oscillating electric field with the same frequency as that of incident IR "light".

All of the motions can be described in terms of two types of molecular vibrations (see Figure 8-1). One type of vibration, a stretch, produces a change of bond length. A stretch is a rhythmic movement along the line between the atoms so that the interatomic distance is either increasing or decreasing.

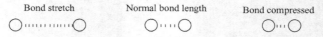

Figure 8-1 Stretching vibration

The second type of vibration, a bend, results in a change in bond angle. These are also sometimes called scissoring, rocking, or "wig wag" motions (see Figure 8-2).

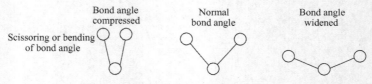

Figure 8-2 Bending vibration

Each of these two main types of vibration can have variations. A stretch can be symmetric or asymmetric. Bending can occur in the plane of the molecule or out of plane; it can be scissoring, like blades of a pair of scissors, or rocking, where two atoms move in the same direction.

Different stretching and bending vibrations can be visualized by considering the CH_2 group in hydrocarbons. The arrows indicate the direction of motion. The stretching motions require more energy than the bending ones (see Figure 8-3).

Note the high wave number (high energy) required to produce these motions.

The bending motions are sometimes described as wagging or scissoring motions (see Figure 8-4).

You can see that the lower wave number values are consistent with lower energy to

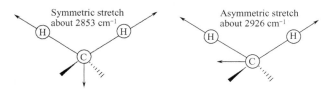

Figure 8-3   Stretching vibration of the CH_2 group hydrocarbons

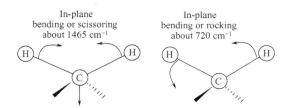

Figure 8-4   bending vibration of the CH_2 group hydrocarbons

cause these vibrations.

A molecule absorbs a unique set of IR light frequencies. Its IR spectrum is often likened to a person's fingerprints. These frequencies match the natural vibrational modes of the molecule. A molecule absorbs only those frequencies of IR light that match vibrations that cause a change in the dipole moment of the molecule. Bonds in symmetric N_2 and H_2 molecules do not absorb IR because stretching does not change the dipole moment, and bending cannot occur with only 2 atoms in the molecule. Any individual bond in an organic molecule with symmetric structures and identical groups at each end of the bond will not absorb the IR range. For example, in ethane, the bond between the carbon atoms does not absorb IR because there is a methyl group at each end of the bond. The C—H bonds within the methyl groups do absorb.

In a complicated molecule many fundamental vibrations are possible, but not all are observed. Some motions do not change the dipole moment for the molecule; some are so much alike that they coalesce into one band.

Even though an IR spectrum is the characteristic for an entire molecule, there are certain groups of atoms in a molecule that give rise to absorption bands at or near the same wave number (frequency), regardless of the rest of the structure of the molecule. These persistent characteristic bands enable you to identify major structural features of the molecule after a quick inspection of the spectrum and the use of a correlation table. The correlation table is a listing of functional groups and their characteristic absorption frequencies.

The infrared spectrum for a molecule is a graphical display. Figure 8-5 shows the frequencies of IR radiation absorbed and the percentage of the incident light that passes through the molecule without being absorbed. The spectrum has two regions. The fingerprint region is unique for a molecule and the functional group region is similar for molecules with the same functional groups.

The nonlinear horizontal axis has units of wave numbers. Each wave number value matches a particular frequency of infrared light. The vertical axis shows the percentage of trans-

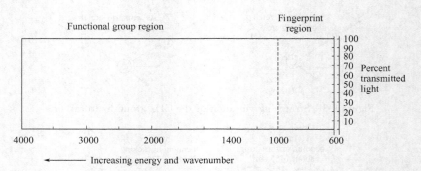

Figure 8-5 Infrared spectrum

mitted light. At each frequency the ％ transmitted light is 100％ for light that passes through the molecule with no interactions; it has a low value when the IR radiation interacts and excites the vibrations in the molecule.

Selected from "Willard, H.; Merritt, L.; Dean, J.A.; Settle Jr., F.A. Instrumental Methods of Analysis, 7th ed., Wadsworth Publishing Co., 1988."

New Words

functional group 官能团
infrared *adj.* 红外线的；*n.* 红外线
spectrum *n.* 光，光谱，型谱
flexible *adj.* 柔韧性，易曲的，灵活的，柔软的，能变形的，可通融的
oscillate *v.* 振荡
bond *n.* 化合键
angle *n.* ［数］角
oscillating electric field 振荡电场
frequency *n.* 频率，周期
stretch *v.* 伸展，伸长；*n.* 伸展
rhythmic movement 韵律活动，律动
interatomic distance 原子间距离
rocking *adj.* 摇摆的，摇动的
symmetric *adj.* 对称的，均衡的
asymmetric *adj.* 不均匀的，不对称的
scissor *vt.* 剪，剪取；*n.* 剪刀
hydrocarbon *n.* 烃，碳氢化合物
wag *vt.* 摇摆，摇动，饶舌
mode *n.* 方式，模式，样式
dipole moment 偶极矩
ethane *n.* 乙烷
methyl group 甲基

nonlinear *adj.* 非线性的
horizontal axis 水平轴（线）
vertical *adj.* 垂直的，直立的

Notes

1. A molecule absorbs infrared radiation when the vibration of the atoms in the molecule produces an oscillating electric field with the same frequency as the frequency of incident IR "light".

    参考译文：当构成分子的原子振动时就产生了和入射的红外光频率完全一致的振荡磁场，这时该分子就吸收了红外光。

2. Bending can occur in the plane of the molecule or out of plane; it can be scissoring, like blades of a pair of scissors, or rocking, where two atoms move in the same direction.

    参考译文：原子振动摇摆时可以在分子的平面内进行，也可以不在分子的平面内进行。它可以做剪切运动，类似剪刀刀刃运动，也可以做摇摆运动，两个原子的运动方向一致。

3. A molecule absorbs only those frequencies of IR light that match vibrations that cause a change in the dipole moment of the molecule.

    参考译文：只有当入射的红外光线的频率与分子的偶极矩运动发生改变而产生的振动频率相匹配时，该分子才能吸收这一频率的红外光线。

4. The vertical axis shows % transmitted light. At each frequency the % transmitted light is 100% for light that passes through the molecule with no interactions; it has a low value when the IR radiation interacts and excites the vibrations in the molecule.

    参考译文：纵坐标表示透光率；在每一频率下，如果入射光透过分子却不与分子相互作用，那么透光率为 100%。如果红外光线与分子相互作用并且引发了分子的振动，那么透光率的值较低。

Exercises

1. Put the following into Chinese.

infrared spectrum	average position
bond angle	infrared radiation
interatomic distance	symmetric
dipole moment	absorption frequency

2. Put the following into English.

有机官能团	键长
频率	烃
有机分子	乙烷
碳原子	甲基

3. Questions.

    (1) What is the method of infrared spectroscopy intended to do?

    (2) Why are bond lengths and bond angles continuously changing?

    (3) How many types of molecular vibrations do you know about?

(4) In which way can we visualize the different stretching and bending vibrations?

(5) Can all the fundamental vibrations be observed in a complicated molecule?

科 技 英 语 基 本 技 巧

翻译技巧：简单定语从句的汉译

所谓简单的定语从句，是指一个句子只含一个定语从句的结构。这类定语从句，不论是限制性的还是非限制性的，在结构上与先行词的关系比较简单，翻译时多译成定语，也可译成并列句、状语从句或独立句等；这种定语从句还可译成句子成分，即和先行词放在一起翻译，它有如下几种情况。

（1）限制性定语从句一般应利用"的"字结构照译为其前词的前置定语。

The polybutylene also has high resistance to corrosive attack by strong acids and bases as is expected for the polyolefine. 聚丁烯也具有聚烯烃可能具有的高的耐强酸强碱腐蚀性。

（2）转译为并列分句。定语从句较长时，也可以用这种方法转译为并列分句。

A multi-purpose machine tool is one, which is capable of doing a number of different types of operations. 多用机床是这样一种机床，即能进行很多种不同类型操作的机床。

（3）改变全句的结构，即将定语从句和主句糅合到一起，译成为一个句子的形式。（该从句在其中充当定语、谓语、主语或状语等）

The yeast produces a mixture of enzymes, which bring about this fermentation. 酵母产生的酶混合物能引起这种发酵作用。（将定语从句转译为谓语）

In the future the energy obtained from uranium atoms in nuclear-power stations may be the main source of power used to heat boilers and produce steam for the turbines that drive the alternators. 将来，在原子能发电厂里从铀原子中获得的能量可能是用于发电的主要能源，这些能源用来加热锅炉，产生蒸汽，以驱动汽轮机，汽轮机再驱动交流发电机。（将定语从句转译为原来主句中的目的状语）

听力技巧：语音方面的难点与策略

形成听力理解困难的原因是多方面的。从语音角度看，在听专业性英语授课、报告或讨论时，应注意以下几个方面。

1. 注意音的连续与失去爆破

语音在语流中会发生某些变化。当两音相邻时，互相影响，有时产生连读，有时则会发生失去爆破。

2. 注意略音

英语口语中由于大量的运用连读、失去爆破、辅音连缀等发音技巧，在快速语流中便很自然地出现不少略音。在专业英语中，这种现象普遍存在。如在下列这个极普通的句子中，就存在连读、失去爆破、辅音连缀等现象。试读这个句子，体会这些发音特点：

As a matter of fact, the electric generator needs another powerful magnets to generate current.

3. 注意重读与弱读

在连贯的语句中，有的词要重读，有的词要轻读。从口语语法上讲，弱读的通常是冠词、介词、连词、助动词、人称代词等功能词。有人统计过，在连续性谈话中有19个常用功能词（例如 at, of, the, to, and, or, a, but, for, then, shall），在90%的场合被一带而过。专业英语中的弱读功能词虽然速度快，不易把握，但就词汇量而言，这些词与实词相比毕竟为数有限。因此应把失去爆破、连读、重读与弱读结合起来练习，试图找出句中重读词与弱读词之间的联系，多听录音，并反复跟读。这样，在听英语课或专业性报告时就不会感到很困难了。

4. 注意偶尔遇到非标准音、语调

有些母语为英语的人因受影响导致元音、辅音误读及任意省略单词中的音素。如有些美国人将 had 读成/had/，把 wild 读成/wail/等。另一方面，使用英语表达的人，其用词并非都是英语，一些非英语国家的人有时讲起英语来就不太规范，甚至还夹杂着其他方言土语。遇到这种情况时，应根据上下文或"听音形"的推测，迅速判断出该词是什么，然后确定出讲话者的发音规律。

Reading Material

Ultraviolet Spectroscopy

Basics of UV Light Absorption

Ultraviolet/visible spectroscopy involves the absorption of ultraviolet/visible light by a molecule causing the promotion of an electron from a ground electronic state to an excited electronic state.

Ultraviolet/Visible light: wavelengths (λ) between 190 nm and 800 nm (see Figure 8-6).

An ultraviolet spectrum is recorded by irradiating the sample with UV light of continuously changing wavelength. When the wavelength corresponds to the energy level required to excite an electron to a higher level, energy is absorbed.

This absorption is detected and displayed on a chart that plots wavelength versus absorbance. UV differs from IR spectra in the way they are presented. IR spectra are usually displayed so that the baseline corresponding to the zero absorption runs across the top of the chart and a valley indicates absorption. UV spectra are displayed with the baseline at the bottom of the chart so that a peak indicates absorption.

Beer-Lambert law

The ultraviolet spectra of compounds are usually obtained by passing light of a given wavelength (monochromatic light) through a dilute solution of the substance in a non-ab-

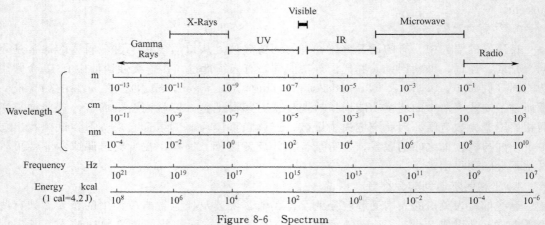

Figure 8-6　Spectrum

sorbing solvent. The intensity of the absorption band is measured by the percent of the incident light that passes through the sample:

$$\% T = (I/I_0) \times 100\%$$

where　I——intensity of transmitted light;

I_0——intensity of incident light.

Because light absorption is a function of the concentration of the absorbing molecules, a more precise way of reporting intensity of absorption is by use of the Beer-Lambert law:

$$A = -\lg(I/I_0) = \varepsilon c l$$

where　ε——molar absorptivity;

c——molar concentration of solute;

l——length of sample cell, cm.

Terminology

The following definitions are useful in a discussion of UV/Vis spectroscopy.

Chromophore: Any group of atoms that absorbs light whether or not a color is thereby produced.

Auxochrome: A group which extends the conjugation of a chromophore by sharing of nonbonding electrons.

Bathochromic shift: The shift of absorption to a longer wavelength.

Hypsochromic shift: The shift of absorption to a shorter wavelength.

Hyperchromic effect: An increase in absorption intensity.

Hypochromic effect: A decrease in absorption intensity.

Measurement of the Spectrum

The UV spectrum is usually taken on a very dilute solution (1 mg in 100 ml of solvent). A portion of this solution is transferred to a silica cell. A matched cell containing pure solvent is prepared, and each cell is placed in the appropriate place in the spectrometer. This is so arranged that two equal beams of light are passed, one through the solution of the sample, one through the pure solvent. The intensities of the transmitted light are then compared over the whole wavelength range of the instrument. The spectrum is plotted automatically as a $\lg 10(I_0/I)$ ordinate and λ abscissa. For publication and comparisons these are often converted to an ε vs. λ or $\lg \varepsilon$ vs. λ plot. The λ unit is almost always in nanometers (nm).

Selected from "Willard, H.; Merritt, L.; Dean, J.A.; Settle Jr., F.A. Instrumental Methods of Analysis, 7th ed., Wadsworth Publishing Co., 1988."

Unit 9

Text: Potentiometric Titration

Description

Titration is a quantitative measuring procedure in which a liquid solution is added to a mixture until some distinctive feature signals an end point. The titration of Fe^{2+} with $KMnO_4$ is studied. As the reaction proceeds, the ratio of Fe^{3+}/Fe^{2+} increases and the electrical potential changes. After the end point, the ratio of MnO_4^-/Mn^{2+} determines the potential. The potential of an electrode placed in this reaction mixture is compared to the potential of a Cu/Cu^{2+} reference system.

Introduction

Suppose an electrochemical cell is created by setting up an experiment according to the Figure 9-1 below.

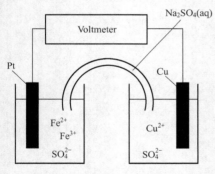

Figure 9-1 Electrochemical cell

This cell corresponds to a chemical reaction that can be written as

$$2Fe^{3+} + Cu \longrightarrow 2Fe^{2+} + Cu^{2+}$$

The voltage read by the voltmeter is a function primarily of the difference in the standard electrode potentials and secondarily of the ratio of the concentrations. What if the ratio of concentrations is changed, say, by oxidizing some of the Fe^{2+} to Fe^{3+} with $KMnO_4$? In this case, the meter reading will change. Measuring changes in these cell potentials is the basis of potentiometric titrations. Once essentially all the Fe^{2+} is reacted, the cell potential is determined by the MnO_4^-/Mn^{2+} couple versus copper.

Safety

The chemicals used are toxic. $KMnO_4$ and H_2SO_4 are corrosive. Wear goggles and apron. Avoid ingesting the chemicals. Avoid contact with the chemicals. Wash spills with water. Wash hands after the experiment.

Procedure

- Place a small number of pieces of dried polyacrylamide gel on to a piece of paper. Use an unbent paper clip to insert a small piece into the end of the tube, and push it about 1.5 cm into the tube.

- Fill a well elsewhere in the plate conveniently removed from the reaction well with 0.1 mol·L^{-1} Na$_2$SO$_4$. Slowly squeeze the bulb, insert it into the Na$_2$SO$_4$ solution, and then release slowly as liquid rises in the plastic tube into the bulb covering the small piece. Once the gel swells, motion of the fluid in the plastic tube is impaired.

- Use a razor to cut a small hole (5~8 mm diameter) in the top of the bulb. The hole must be smaller than the diameter of your copper coil. Use a transfer pipet to remove excess Na$_2$SO$_4$. Take a 12~15 cm length of solid copper wire and form a coil around a pen or pencil.

- Fill the bulb 2/3rds full with 0.5 mol·L^{-1} CuSO$_4$. This apparatus is the reference cell. Put a small amount of water in one well. Between titrations, store this reference cell with the tip down in the water.

- Clamp one end of a multimeter to the copper wire protruding from the bulb of the copper sulfate copper reference cell. Place a piece of platinum wire in one well of a 24-well plate. Nichrome wire is a substitute; it reacts at the end, however. Also, it must be pretreated. Bend the wire so that it fits snugly in the center of the cell. Clamp the other lead of the multimeter to the wire.

- Switch the meter to a voltage scale. Turn it on. The reading should be 0.00 mV, or just a few mV (neglect sign). Trapped air bubbles defeat the purpose of the bridge. Place the tip of the bulb of reference cell into the reaction cell. The meter should now give a reading.

- Add 20 drops of 0.2 mol·L^{-1} Fe(NH$_4$)$_2$(SO$_4$)$_2$ to the platinum electrode well (the reaction well). Add 5 drops of 3 mol·L^{-1} H$_2$SO$_4$ to the well with the platinum electrode. Stir using the reference cell. Note and record the meter reading (It may be unstable at this point.). Add 1 drop of 0.050 mol·L^{-1} KMnO$_4$ to the reaction well. Stir using the ref-erence cell. When the reading becomes stable, record the reading. Note and record any evidence for reaction. Continue the process of adding one drop of KMnO$_4$, stirring, waiting for a (fairly) stable reading, and recording. Note and record any evidence for reaction.

- Continue adding KMnO$_4$ dropwise, with stirring, until the cell turns pinkish. Then add dropwise 5 drops of KMnO$_4$. After each drop, stir, record the voltage, and record observations.

- Disconnect the meter. Remove the reference cell. Remove the electrode from the reaction well and wash the electrode with a stream of distilled water. Discard the contents of the reaction wells into a disposal beaker provided by the instructor (Use a transfer pipet to suck out the contents from the well and transfer them to the disposal beaker or jar. Then rinse the entire plate at the sink).

Selected from "D. H. Geske and A. J. Bard," Evaluation of the Effect of Secondary Reactions in Controlled Potential Coulometry, "J. Phys. Chem., 63, 1057 (1959)."

New Words

distinctive *adj.* 与众不同的，有特色的
electrical potential 电动势

ratio   *n*. 比，比率
electrode   *n*. 电极
electrochemical cell   电化电池
chemical reaction   化学反应
voltage   *n*. ［电工］电压，伏特数
voltmeter   *n*. 伏特计
copper   *n*. 铜
toxic   *adj*. 有毒的，中毒的
corrosive   *adj*. 腐蚀的，蚀坏的，有腐蚀性的；*n*. 腐蚀物，腐蚀剂
goggle   *n*. 护目镜
ingest   *vt*. 摄取，咽下，吸收
spill   *n*. 溢出，溅出；*vt*. 使溢出
polyacrylamide   *n*. ［化］聚丙烯酰胺
gel   *n*. 凝胶体
tube   *n*. 管，管子
squeeze   *n*. 压榨，挤；*v*. 压榨，挤，挤榨
bulb   *n*. 球形物
fluid   *n*. 流动性，流度；*adj*. 流动的
reference cell   参比池
multimeter   *n*. 万用表
sulfate   *n*. ［化］硫酸盐
platinum wire   铂丝，白金丝
nichrome wire   镍铬合金线，镍铬电热丝
scale   *n*. 刻度，衡量，比例

Notes

1. Titration is a quantitative measuring procedure in which a liquid solution is added to a mixture until some distinctive feature signals an end point.

   参考译文：滴定是一种定量测定过程，在这个过程中，将一种溶液滴加到某一种混合物之中直到出现明显现象标志着滴定终点的到达。

2. The voltage read by the voltmeter is a function primarily of the difference in the standard electrode potentials and secondarily of the ratio of the concentrations. What if the ratio of concentrations is changed, say by oxidizing some of the Fe^{2+} to Fe^{3+} with $KMnO_4$?

   参考译文：伏特计上读取的电压主要取决于标准电极电势的差额，其次取决于离子浓度的比值。假如高锰酸钾将 Fe^{2+} 氧化成 Fe^{3+}，离子浓度的比值发生变化，电压会怎么变化？

3. Slowly squeeze the bulb, insert it into the Na_2SO_4 solution, and then release slowly as liquid rises in the plastic tube into the bulb covering the small piece. Once the gel swells, motion of the fluid in the plastic tube is impaired.

   参考译文：慢慢地挤压吸耳球，将其插入 Na_2SO_4 溶液，然后慢慢地放松吸耳球同时塑料管中的液体会提升到标线处。一旦橡胶体膨胀，塑料管中的液体的运动就会削弱。

PART 2　INSTRUMENT ANALYSIS

Exercises

1. Put the following into Chinese.

   quantitative measure　　　　　　　electrical potential
   polyacrylamide　　　　　　　　　platinum
   multimeter　　　　　　　　　　　nichrome
   platinum electrode

2. Put the following into English.

   电位滴定法　　　　　　　　　　　化学反应
   伏特计　　　　　　　　　　　　　标准电极电势
   铜　　　　　　　　　　　　　　　参比池
   硫酸铜

3. Questions.

   (1) How many steps do we have to follow in the procedure of potentiometric titration?

   (2) Do you know how to create an electrochemical cell?

   (3) Can you illustrate the chemical reaction $2Fe^{3+} + Cu \longrightarrow 2Fe^{2+} + Cu^{2+}$ titration in our simple English?

   (4) For the sake of safety, what will be done before and after the potentiometric titration?

   (5) What is the definition of titration?

科 技 英 语 基 本 技 巧

翻译技巧：状语从句的译法

英语状语从句表示时间、原因、条件、让步、目的等，英语状语从句用在主句后面的较多，而汉语的状语从句用在主句前的较多，因此，在许多情况下，应将状语从句放在主句前面。下面我们通过一些实例说明它们常用的翻译方法。

(1) For example, when steam passes over red-hot iron, iron oxide and hydrogen form. 当蒸汽从炽热的铁上通过时，就会产生氧化铁和氢气。（时间状语从句）

(2) The selectivity of the electrode increases as the concentration of organic salts decreases. 当有机盐的浓度降低时，电极的选择性就增大。（时间状语从句）

(3) Some sulphur dioxide is liberated when coal, heavy oil and gas burn, because they all have certain sulphur compounds. 因为煤、重油和煤气都含有硫化物，所以它们燃烧时会放出一些二氧化硫。（原因状语从句）

(4) The induced EMF is in such a direction that it opposes the change of current. 感应电动势的方向是阻止电流发生变化的那个方向。（将结果状语从句译成表语从句）

(5) Should something go wrong, the control rod would drop. 万一发生什么事故，控制棒就会掉下来。（条件状语从句）

(6) Iron products are often coated lest they should rust. 铁制品常常涂以保护层，以免生锈。（目的状语从句）

(7) Though the cost of the venture would be immense, both in labour and power, many believe that iceberg towing would prove less costly in the long run than the alternative of desalination of sea water. 这种冒险的代价，不管是在人力还是在能源消耗方面，都将是巨大的。然而，许多人认为，冰山牵引最终会证明比选择海水脱盐法花费要少。（让步状语从句）

听力技巧：语法方面的难点与策略

无论是专业报告、专业论文还是专业课程讲义，其语法句式结构都有其特点和难点。在掌握这些特定语法结构的基础上来听英语讲课或演讲，会有事半功倍的效果。

1. 陈述句多

在专业英语口语中，当描述实验、说明现象、明确定义、表达定理、定律和原理时，多用陈述语气，很少使用疑问句，几乎不用感叹句。例如：

Here a tension load F is applied through pins at the end of the bar.

2. 祈使句多

科学研究领域中的一个重要标准就是所述理论具有可解释性，所做实验具有可重复性。无论谁照此去做都是可以的，而且由于讲话人与听众属于直接交流，因此常常没有必要指明主语。例如：

Let's look at the flask again. Put 100 grams of water in the beaker.

3. 简单句多

人们在日常交谈、授课乃至专业性演讲时经常使用简单句以起到简洁明了的效果。例如：

Electricity does a lot of things for us. Filing should be avoided except to break sharp corners.

4. 被动语态多

专业英语描述客观事实，一般着重强调是的所涉及的事物，要求见物、见动作，尽量避免个人感觉和判断，因此大量采用以客观事物为主体的被动语态。例如：

The storage battery is being charged.

5. 虚拟语气多

在专业英语口语中，讲话人在表示自己对各种问题的看法时，为了使语气委婉，往往采用虚拟语气。例如：

For example, it would be better to use 2-volt cells in parallel.

Reading Material

Principles of the Glass Electrode Method

In the glass electrode method, the known pH of a reference solution is determined by using two electrodes, a glass electrode and a reference electrode, and measuring the voltage (difference in potential) generated between the two electrodes. The difference in pH between solutions inside and outside the thin glass membrane creates electromotive force in proportion to this difference in pH. This thin membrane is called the electrode membrane. Normally, when the temperature of the solution is 30 ℃, if the pH inside is different from that of outside by 1, it will create approximately 60 mV of electromotive force.

The liquid inside the glass electrode usually has a pH of 7. Thus, if one measures the electromotive force generated at the electrode membrane, the pH of the test solution can be found by calculation.

A second electrode is necessary when you are measuring the electromotive force generated at the electrode membrane of a glass electrode. This other electrode, paired with the glass electrode, is called the reference electrode. The reference electrode must have extremely stable potential. Therefore, it is provided with a pinhole or a ceramic material at the liquid junction.

In other words, a glass electrode is devised to generate accurate electromotive force due to the difference in pH (see Figure 9-2). And a reference electrode is devised not to cause electromotive force due to a difference in pH.

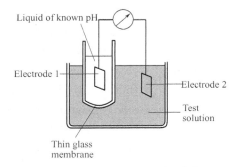

Figure 9-2 Operating principle of the glass electrode method

Explanation of Part of a Glass Electrode pH Meter

A glass electrode pH meter consists of a detector, an indicator and reference solution. A brief description of each part follows.

1. Detector

The detector consists of a glass electrode, a reference electrode, and a temperature-compensation electrode. There is also a composite electrode, in which the glass electrode and the reference electrode are integrated into one unit, and the integrated electrode, into which all three of the above-mentioned electrodes are integrated into a single unit.

2. Glass Electrode

A glass electrode consists of an electrode membrane that responds to pH, a highly isolating base material to support the unit, solution inside the glass electrode, an internal electrode, a lead wire, and a glass electrode terminal.

The most critical item in this system is the electrode membrane. First, the membrane glass must generate a potential that accurately corresponds to the pH of the solution. Second, even though it must be accurately sensitive to acidity and alkalinity, it must not be damaged by them. Third, the electric resistance of the membrane itself must not be too large. Fourth, too large a difference in potential (asymmetric difference in potential) must not be generated between the solutions inside and outside the electrode when the electrode is immersed in a solution of identical pH to that of the solution inside of the electrode. Another requirement is that the glass membrane be resistant to shock and chemical reactions.

Generally, silver chloride is used as the material for the internal electrode. Potassium chloride solution maintained at pH 7 is usually used as the internal solution.

Birth and History Glass Electrode

In 1906, Cremer blew the tip of a glass tube into a bubble and measured the difference in potential between two kinds of solutions (0.6% NaCl+diluted H_2SO_4 and 0.6% NaCl+diluted NaOH). This is considered the birth of the glass electrode. In 1909, Habert and Klemensiewicz measured the difference in potential between a silver chloride electrode and a mercurous chloride electrode, and found that they could obtain a titration curve similar to that of a hydrogen electrode. They called this a glass electrode. So, the glass electrode took its first step toward becoming a practical pH electrode. However, early glass electrodes had large electrical resistance and very thin glass membranes. Therefore, they were very fragile and difficult to handle.

Later, with the introduction of glass containing lithium, which is chemically strong and has low electric resistance and with development of technology for fabricating electronic parts and insulation materials, the glass electrode made rapid progress after the Second World War. Now it is widely used as the standard for measuring pH.

In Japan, professor Tatsuzo Okada of Kyoto University launched a study on lithium glass electrodes right after the end of the war. Also, studies on reference electrodes and amplifiers were carried out by people in various fields. Horiba Radio Laboratory (the predecessor of Horiba, Ltd.) introduced and integrated these technologies and developed the first glass electrode pH meter in Japan in 1950.

Moreover, Horiba introduced a two-dimensional processing technique in creating the structure for the glass electrode and succeeded in the development of the "sheet-type composite glass electrode," which laces the glass electrode and reference electrode, and is only 1 mm in thickness.

Selected from "A. J. Bard," Effect of Electrode Configuration and Transition Time in Solid Electrode Chronopotentiometry, "Anal. Chem., 33, 11 (1961)."

Unit 10

Text: Atomic Absorption Spectroscopy

The popularity of this technique has increased in recent years due to decreased limits of detection. They have moved from 1 ppm (parts per million) to around 100 ppb (parts per billion). The flame technique uses an air-acetylene flame to reach temperatures usually reported around 2360 K and 2600 K, which is used to produce a good atom cell across which the absorption can be measured. Some 30 or 40 elements can be determined in the ppm range with this flame.

In atomic absorption, the atoms absorb part of the light from the source and the remainder of the light not absorbed, P, reaches the detector. Below Figure 10-1 is a block diagram of the major components of an atomic absorption spectrometer.

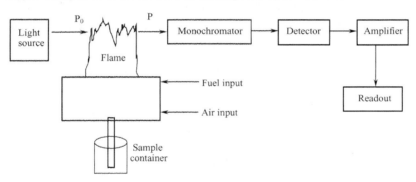

Figure 10-1　Major components of an atomic absorption spectrometer

Liquid sample is aspirated (sucked) into the flame whose temperature ranges from 2000~3000 K depending on flame profile. The liquid evaporates and the remaining solid is atomized (broken into atoms) in the flame. The light source consists of a hollow-cathode lamp (HCL).

A HCL is filled with Ne or Ar gas at a pressure of 1~5 torr. The hollow cathode is coated or filled with the element to be analyzed. This means that if the concentration of calcium in drinking water is to be analyzed, the hollow cathode would either be coated or filled with elemental calcium. Application of a high voltage (600~1000 V) between the anode and cathode ionizes the filler gas, i. e. sufficient energy is absorbed by the filler gas to dislodge an outer electron from a neutral Ne or Ar atom.

$$Ar(g)+e \longrightarrow Ar^+(g)+2e$$
$$Ne(g)+e \longrightarrow Ne^+(g)+2e$$

The $Ar^+(g)$ or $Ne^+(g)$ cations are attracted to the negatively charged cathode. The gaseous ions accelerate towards the cathode and subsequently smash into it with enough kinetic energy to knock the metal atoms from the cathode surface into the gas phase.

The free metal gaseous atoms smash into one another and also with the high-energy electrons released from the ionization of the filler gas. The gaseous metal atoms absorb the kinetic energy from the high-energy electrons, resulting in the excitation of electrons in the gaseous metal to a higher energy level (M^*) where, after a short time, the energetically excited electron emits a photon of energy unique to the metal as it returns to the ground state energy level. From there the process begins anew. M=metal of hollow cathode.

$$M(g) + energy \longrightarrow M^*(g) \quad \text{absorption of energy}$$
$$M^*(g) \longrightarrow M(g) + photon \quad \text{emission of energy}$$

Flame temperature determines the degree to which an analyte sample breaks down to atoms and the extent to which a given atom is found in its ground, excited, or ionized energy state. Each of these effects influences the strength of the signal observed by the spectrometer detector. But not all flames are equal. Furthermore, there are vertical temperature differences within a single flame. One can easily see this by observing the different colored regions of a candle or Bunsen burner flame. Some regions of a candle flame are more orange or yellow than other regions and in a Bunsen burner flame more blue than other regions.

Important regions of a flame used in atomic spectroscopy are shown below Figure 10-2.

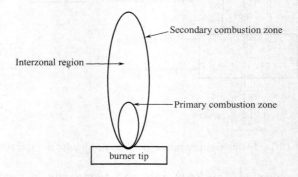

Figure 10-2 Regions of a flame

The appearance and relative size of these three regions vary considerably with the fuel-oxidant ratio (lean versus rich) as well as with fuel (natural gas, molecular hydrogen, acetylene, and nitrous oxide) and oxidant (air or 100% molecular oxygen) type.

Selected from "Willard, H.; Merritt, L.; Dean, J.A.; Settle Jr., F.A. Instrumental Methods of Analysis, 7th ed., Wadsworth Publishing Co., 1988."

New Words

popularity   *n.* 普及，流行，声望
acetylene   *n.* [化] 乙炔，电石气
detector   *n.* 检测器，检波器
monochromator   *n.* 单色器

element　　*n*. 元素
aspirate　　*v*. 吸气
profile　　*n*. 剖面，侧面，外形，轮廓
atomize　　*vt*. 使分裂为原子，将……喷成雾状
hollow cathode lamp（HCL）　　空心阴极灯
torr　　*n*. 托（真空度单位）
anode　　*n*. ［电］阳极，正极
ionize　　*vt*. 使离子化；*vi*. 电离
negatively　　*adv*. 否定地，消极地
energy state　　能态
kinetic energy　　动能
free metal　　游离金属
photon　　*n*. ［物］光子
ground state　　基态
spectrometer　　*n*. ［物］分光计
Bunsen burner　　本生灯（即煤气灯）
nitrous oxide　　［化］一氧化二氮，笑气（laughing gas）

Notes

1. The flame atomization technique uses an air-acetylene flame to reach temperatures usually reported around 2360 K and 2600 K, which is used to produce a good atom cell across which the absorption can be measured.

    参考译文：据报道，空气-乙炔火焰通常可以达到2360～2600 K的高温，火焰原子化技术利用这一特性，使用乙炔气体作为原子吸收池，通过该吸收池便能测定出吸收值。

2. In atomic absorption, the atoms absorb part of the light from the source and the remainder of the light not absorbed, reaches the detector. Below is a block diagram of the major components of an atomic absorption spectrometer.

    参考译文：在原子吸收中，原子从光源吸收部分光线，未被吸收的透射光线到达检测器上。下面是原子吸收分光光度计主要部件的简图。

3. Application of a high voltage (600～1000 V) between the anode and cathode ionizes the filler gas, i. e. sufficient energy is absorbed by the filler gas to dislodge an outer electron from a neutral Ne or Ar atom.

    参考译文：在阳极和阴极之间使用600～1000V的高压可以使输入的气体电离，即输入的气体可以吸收足够的能量来驱走中性的氖原子或氩原子的外层电子。

4. The gaseous ions accelerate towards the cathode and subsequently smash into it with enough kinetic energy to knock the metal atoms from the cathode surface into the gas phase.

    参考译文：气体离子快速流向阴极，于是有足够的动能撞击金属原子，使阴极表面的金属原子剥离进入气相之中。

5. The gaseous metal atoms absorb the kinetic energy from the high-energy electrons, re-

sulting in the excitation of electrons in the gaseous metal to a higher energy level (M^*) where, after a short time, the energetically excited electron emits a photon of energy unique to the metal as it returns to the ground state energy level. From there the process begins anew.

参考译文：气态金属离子吸收了高能电子的动能，使气态金属的电子激发到高能的激发态。达到该能态之后不久，被激活了的电子释放出该金属特有的光子能量同时返回基态能量状态。从基态能量又开始重新循环该过程。

Exercises

1. Put the following into Chinese.

   limits of detection   acetylene flame
   gaseous ion   free metal
   kinetic energy   fuel-oxidant ratio

2. Put the following into English.

   检测器   原子吸收光谱法
   元素   空心阴极灯
   分光计   一氧化二氮

3. Questions.

   (1) Why is atomic absorption spectroscopy becoming more and more popular in recent years?

   (2) How many elements can be determined in the ppm rang with an air-acetylene flame?

   (3) What is the reported temperature suitable in an atomic absorption spectroscopy?

   (4) Sometimes liquid sample can be sucked into the flame at 2000 K, while at the other times it can be sucked into the flame at 3000 K, why?

Lab Key

科 技 英 语 基 本 技 巧

翻译技巧：名词性从句的译法

英语名词性从句包括主语从句、宾语从句、表语从句和同位语从句，在翻译这类从句时，大多数可以按照原文的句序翻译成相应的汉语，但是也有一些具体的处理方法，下面结合一些实例加以说明。

1. 主语从句

(1) 引导词 it 作为主句的形式主语，而将主语从句（其前加连词）放在主句的后面。对这种句型大都可以顺译为无人称句。例如

It is likely that from now on increasing use will have to be made of the heavier petroleum fractions which are, at the moment, less urgently in demand. 很可能从现在开始将不得

不逐步增加目前还不迫切需要的重油馏分的使用量。

（2）有关连词 what、which、how、why、when、where、who、whatever、whoever、whenever、wherever 及从属连词 that、whether、if 等引导出主语从句。它们一般是照译在句首，并用作整个主从复合句的主语。例如

What Faraday called the "electrochemical equivalent" is what we now called the chemical equivalent. 法拉第当年称为"电化学当量"的，就是现在所说的化学当量。

2. 宾语从句

宾语从句可照原文顺序译成汉语。例如

One wonders why a hard diamond can be converted to soft graphite or why table salt dissolves in water and diamonds do not. 人们感到奇怪的是为什么硬的金刚石会变成软的石墨，或者为什么食盐溶于水，而金刚石不溶于水。

3. 表语从句

表语从句是位于主句的连系动词后面、充当主句主语的表语的从句。它是由连词和关联词引导的。和宾语从句一样，表语从句采用顺译法，即先译主句，后译从句。

An even more possibility is that the dark matter is composed of sub-atomic articles left over the Big Bang that is believed to have sparked the creation of the universe. 另外一种可能是：那些黑色物体是由"大碰撞"遗留的亚原子微粒组成。据认为"大碰撞"创造了宇宙。（表语从句）

4. 同位语从句

同位语从句可以根据具体情况译成名词从句、定语从句或外位成分。

These uses are based on the fact that silicon is a semiconductor of electricity. 这些用途是基于硅是半导体这一事实。

The operation is adaptive in the sense that the tap-gain-adjustment information is derived from the received data. 抽头-增益-调节的信息是根据所接受的数据推导出来，在此意义上说，工作是自适应的。（注意本句是将同位语从句译成独立的分句，而后在译出相关词及主句；就中文来讲，前面的分句是外位成分，和"意义"二字同位）

听力技巧：词汇方面的难点与策略

词汇无论在英语学习中还是在使用中都起着重要作用。如果词汇掌握不好，人们就不能将听到的音同其代表的形与义联系起来，也就无法领会其含义。因此，在听专业英语讲课或报告前应注意以下两方面。

1. 专业词汇与普通词汇"两手抓"

专业英语中有不少专业词汇。如果对这些专业词汇不熟悉，将会对听力理解造成很大障碍。另一方面，专业人员在日常交流中并非满口纯专业术语，相反，更多的还是普通词汇。有人曾对一位计算机讲师的授课过程作过统计，20分钟内用词3309个，专业语汇为953个，仅占总数的30%还弱。因此，要提高听力技能，还应掌握大量的普通词汇。

2. 注意一词多义现象

许多人在听英语授课或演讲时，常常会遇到"听清了"但"没听懂"这样的情况。造成此种情形的原因很大程度上是由于英语中一词多义、一义多词的现象很普遍。do, make,

> take 这类普通动词在口语中，其词义为数很多。像 body，work 这样一些很普通的名词，用在不同的专业中，所表示的概念也不同。例如：work 在金属工艺方面，表示"工件"、"工作物"、"加工"、"作业"、"工作"；在机械工程方面，表示"结构"、"机构"、"构件"；在电学方面，则表示"工作"、"制品"等。

Reading Material

Atomic Absorption Spectrophotometer

Job of the Hollow Cathode Lamp
- Provide the analytical light line for the element of interest
- Provide a constant yet intense beam of that analytical line

Job of the Nebulizer
- Suck up liquid sample at a controlled rate
- Create a fine aerosol for introduction into the flame
- Mix the aerosol and fuel and oxidant thoroughly for introduction into the flame

Job of the Fame
- Destroy any analyte ions and breakdown complexes
- Create atoms (the elemental form) of the element of interest Fe, Cu, Zn, etc.

Job of the Monochromator
- Isolate analytical lines' photons passing through the flame
- Remove scattered light of other wavelengths from the flame
- In doing this, only a narrow spectral line impinges on the PMT

Job of the Photomultiplier Tube (PMT)
- As the detector the PMT determines the intensity of photons of the analytical line exiting the monochromator.

The Hollow Cathode Lamp

The hollow cathode lamp (HCL) uses a cathode made of the element of interest with a low internal pressure of an inert gas. A low electrical current (about 10 mA) is imposed in such a way that the metal is excited and emits a few spectral lines characteristic of that element (for instance, Cu 324.7 nm and a couple of other lines; Se 196 nm and other lines, etc.). The light is emitted directionally through the lamp's window, a window made of a glass transparent in the UV and visible wavelengths.

The nebulizer chamber thoroughly mixes acetylene (the fuel) and oxidant (air or nitrous oxide), and by doing so, creates a negative pressure at the end of the small diameter, plastic nebulizer tube. This negative pressure acts to suck ("uptake") liquid sample up the tube and into the nebulizer chamber, a process called aspiration. A small glass impact bead and/or a fixed impeller inside the chamber creates a heterogeneous mixture of gases (fuel+oxidant) and suspended aerosol (finely dispersed sample). This mixture flows immediately into the burner head where it burns as a smooth, laminar flame evenly distributed along a narrow slot in the well-machined metal burner head.

Liquid sample not flowing into the flame collects on the bottom of the nebulizer chamber and flows by gravity through a waste tube to a glass waste container (remember, this is still highly acidic).

For some elements that form refractory oxides (molecules hard to break down in the flame) nitrous oxide (N_2O) needs to be used instead of air (78% N_2 + 21% O_2) for the oxidant. In that case, a slightly different burner head with a shorter burner slot length is used.

Tuned to a specific wavelength and with a specified slit width chosen, the monochromator isolates the hollow cathode lamp's analytical line. Since the basis for the AAS process is atomic absorption, the monochromator seeks to only allow the light not absorbed by the analyte atoms in the flame to reach the PMT. That is, before an analyte is aspirated, a measured signal is generated by the PMT as light from the HCL passes through the flame. When analyte atoms are present in the flame——while the sample is aspirated——some of that light is absorbed by those atoms (remember it is not the ionic but elemental form that absorbs). This causes a decrease in PMT signal that is proportional to the amount of analyte. This last is true inside the linear range for that element using that slit and that analytical line. The signal is therefore a decrease in measure light: atomic absorption spectroscopy.

State of Samples and Standards

The samples and standards are often prepared with duplicate acid concentrations to replicate the analyte's chemical matrix. closely as possible. Acid contents of 1%~10% are common. In addition, high acid concentrations help keep all dissolved ions insolution.

Ignition, Flame Conditions, and Shut down

The process of lighting the AAS flame involves turning on first the fuel then the oxidant and then lighting the flame with the instrument's auto ignition system (a small flame or red-hot glow plug). After only a few minutes the flame is stable. Deionized water or a dilute acid solution can be aspirated between samples. An aqueous solution with the correct amount of acid and no analyte is often used as the blank.

Careful control of the fuel/air mixture is important because each element's response depends on that mixture in the burning flame. Remember that the flame must breakdown the analyte's matrix and reproducibly create the elemental form of the analyte atom. Optimization is accomplished by aspirating a solution containing the element (with analyte content about that of the middle of the linear response range) and then adjusting the fuel/oxidant mix until the maximum light absorbance is achieved. Also the position of the burned head and nebulizer uptake rate are similarly "tuned". Most computer controlled systems can save variable settings so that methods for different elements can be easily saved and reloaded.

Shut down involves aspirating deionized water for a short period and then closing the fuel off first. Most modern instruments control the ignition and shutdown procedures automatically.

Selected from "Schoeff, M. S. and Williams, R. H. Principles of Laboratory Instruments, Mosby: St. Louis, 1993."

Unit 11

Text: Gas Chromatography

Description

Chromatography is the science of separation, which uses a diverse group of methods to separate closely related components of complex mixtures. During gas chromatographic separation, the sample is transported via an inert gas called the mobile phase. The mobile phase carries the sample through a coiled tubular column where analytes interact with a material called the stationary phase. For separation to occur, the stationary phase must have an affinity for the analytes in the sample mixture. The mobile phase, in contrast with the stationary phase, is inert and does not interact chemically with the analytes. The only function of the mobile phase is to sweep the analyte mixture through the length of the column. Gas chromatography can be divided into two categories: (1) gas-solid and (2) gas-liquid chromatography. Gas-liquid GC, developed in 1941, is the primary GC technique used for environmental applications. Gas-solid GC is not widely used for environmental applications.

The stationary phase is chosen so that the components of the sample distribute themselves between the mobile and stationary phase to varying degrees. Those components that are strongly retained by the stationary phase move slowly relative to the flow of the mobile phase. In contrast, components that have a lower affinity for the stationary phase travel through the column at a faster rate. As consequence of the differences in mobility, sample components separate into discrete bands that can be analyzed qualitatively and quantitatively.

Gas chromatography is the most widely used chromatographic technique for environmental analyses.

System Components

The primary components of a GC include:
- injection port
- column
- integrator or data acquisition system
- detectors

Other parts include:
- autosampler (s)
- control panel, electronic pressure control (EPC)
- injection port liners
- septa
- ferrules
- flow controllers

The carrier gas is introduced in the injection port where the sample is volatilized and swept through the column, and where the compounds are separated. The carrier gas/sample mixture then enters the detector where the compounds are identified. The signal from the detector then is amplified and displayed by the data system.

A capillary column is an open tube made of fused silica with an outer coating of durable plastic and an inner coating of stationary-phase material. Some capillary columns have a second outer covering of stainless steel to withstand the higher pressure required to analyze alcohols, ketone, and VOCs by the purge-and-trap method. A lesser used column type is the packed column. Packed columns use a stainless steel or glass tube with a 1/8th inch inner diameter packed with a solid stationary phase.

The effectiveness of a chromatographic column in separating solutes (analytes) is dependent on a number of variables. Understanding these variables is essential to the process of optimizing any chromatographic system and achieving resolution of analytes. Variables that affect separation include distribution equilibrium constants, retention time, retention (capacity) factors, and selectivity factors.

Theory of Operation

The theory of separation by GC is relatively simple and understanding the factors that affect separation allows more effective applications of GC analysis in the field. The purpose of separation is to allow identification and quantitation of individual components of a mixture and the theory of separation is detailed below. In addition to separation, detection of analytes after separation, which is an essential but separate aspect of chromatography, is presented in the section describing system components. Basic components of a complete gas chromatographic system include:

- a carrier gas supply
- a syringe for sample introduction
- the injection port
- the column and oven
- the detector and data collection system

Components of a gas chromatograph are presented in greater detail in the section describing the system components. Schematic diagrams and photographs of instruments can also be accessed through the systems components section.

Before separation occurs in the chromatographic column, the mixture of components in the sample is introduced into the chromatograph through the injection port with a syringe. At this point, the analytes are vaporized (if not already in the gas phase) by the high temperature maintained in the injection port. The analytes are kept in the gaseous state by maintaining all elements of the instrument at a temperature above the boiling point of the analytes. The gas phase analytes are then immediately swept onto the chromatographic column by the mobile phase. The mobile phase is comprised of an inert carrier gas, which usually is nitrogen, helium, or hydrogen.

As the analytes are swept through the column by the mobile phase, separation occurs based on the affinity of each analyte for the stationary phase. The gas chromatographic column is composed of a coiled, tubular column and the stationary phase within the tube. GC columns are either packed or open-tubular. Early GC columns were packed with carbon or diatomaceous earth based solids which acted as the stationary phase. In modern open-tubular columns, the stationary phase is a liquid organic compound that is coated on the internal surface of the fused silica column. Polarities of the analytes dictate the choice of stationary phase. Components of the mixture with a high degree of affinity for the stationary phase are strongly retained while components with low affinity for the stationary phase migrate rapidly through the column. As a consequence of the differences in mobility due to affinities for the stationary phase, sample components separate into discrete band that can be qualitatively and quantitatively analyzed.

As individual components of the mixture elute the chromatographic column, they are swept by the carrier gas to a detector. The detector generates a measurable electrical signal, referred to as peaks, that is proportional to the amount of analyte present. Detector response is plotted as a function of the time required for the analyte to elute from the column after injection. The resulting plot is called a chromatogram. Detector response is generally a gaussian shaped curve representative of the concentration distribution of the analyte band as it elutes from the column. The position of the peaks on the time axis may serve to identify the components and the area under the peaks provide a quantitative measure of the amount of each component.

Selected from "Schoeff, M. S. and Williams, R. H. Principles of Laboratory Instruments, Mosby: St. Louis, 1993."

New Words

chromatography   *n.* 色谱，色谱法
inert gas   惰性气体
mobile phase   （色谱分析用的）流动相
coil   *v.* 盘绕，卷
tubular   *adj.* 管状的
stationary phase   （色谱分析用的）固定相
affinity   *n.* 吸引力，亲和力
column   *n.* 圆柱，柱状物，色谱柱
distribute   *vt.* 分发，分配，散布；*v.* 分发
injection   *n.* 注射，注射剂
control panel   ［计］控制面板
carrier gas   载气
amplify   *vt.* 放大，增强；*v.* 扩大
capillary   *n.* 毛细管；*adj.* 毛状的，毛细作用的

fused silica 熔融石英
stainless steel 不锈钢
alcohol *n*. 酒精，酒
ketone *n*. ［化］酮
retention time 保留时间
vaporize *v*. （使）蒸发
helium *n*. 氦（元素符号 He）
elute *vt*. ［化］洗提，洗脱

Notes

1. Chromatography is the science of separation which uses a diverse group of methods to separate closely related components of complex mixtures.

    参考译文：色谱法是一种分离技术，该分离技术使用各种方法来使复杂的混合试样中化学性质相似的组分分离开来。

2. During gas chromatographic separation, the sample is transported via a gas called the mobile phase. The mobile phase carries the sample through a coiled tubular column where analytes interact with a material called the stationary phase.

    参考译文：在气相色谱分离过程中，我们使用气体作为流动相，流动相载着试样经过一个螺旋的管状色谱柱，在色谱柱中被测物质与固定相相互作用。

3. Gas chromatography can be divided into two categories：（1）gas-solid and （2）gas-liquid chromatography. Gas-liquid GC, developed in 1941, is the primary GC technique used for environmental applications. Gas-solid GC is not widely used for environmental applications.

    参考译文：气相色谱法可划分为两大类：（1）气固色谱法和（2）气液色谱法。气液色谱法是在1941年发展起来的，它主要是用于环保方面的色谱法。气固色谱法在环保方面应用并不广泛。

4. Those components that are strongly retained by the stationary phase move slowly relative to the flow of the mobile phase. In contrast, components that have a lower affinity for the stationary phase travel through the column at a faster rate.

    参考译文：相对于流动相而言，那些被固定相强有力地保留下来的组分的流速较慢。与此相反，那些和固定相的作用力不大的组分流经色谱柱时流速较快。

5. Variables that affect separation include distribution equilibrium constants, retention time, retention (capacity) factors, and selectivity factors.

    参考译文：影响分离的变量有分配系数、保留时间、容量因子和选择因子。

Exercises

1. Put the following into Chinese.

    gas-liquid chromatography   stationary phase
    qualitative and quantitative analysis   injection port
    capillary column

2. Put the following into English.

惰性气体　　　　　　　　　　　　　　流动相
保留时间　　　　　　　　　　　　　　酒精
载气

3. Questions.

(1) What is the most commonly used chromatographic technique for environmental analyses?

(2) How many primary components of a GC include?

(3) What are the deciding variables in the effectiveness of a chromatographic column?

(4) What are the two categories into which gas chromatography is divided?

(5) What is the gas chromatography column?

Lab Key

科技英语基本技巧

听力技巧：听力与其他技能的关系

1. 听与写并举

一些人在听专业性讲课或演讲时，能听懂所听内容的每一句话，但听完之后整体印象却不深，或者基本概念还模糊，或者主要信息没有把握住。其原因之一就是没养成边听边记录的习惯。因此要想不忘记所听到的内容，就要注意听与写的并举，养成理解记忆的习惯。记录的内容可以是关键词，也可以是数字或记忆符号之类的信息。记录速度要快，并以不影响听下面的内容为原则。

2. 听与说并举

在学习英语的过程中，常常会出现这样一些现象：听懂的东西未必都能说得出，但说得出的东西通常都能听懂。这是因为说得出的东西在头脑中的语言记忆一般都比较深刻。所以除了平时加以模仿，提高跟读、朗读和用英语进行口头交流的能力外，在听英语授课或演讲时，如果遇到新的单词或语法结构，可私下说几遍以加强记忆，以便以后再听到时能迅速听懂。

3. 听与听写并举

无论是听专业英语讲课还是听专业英语报告，都要做记录。其实边听边记录在某种程度上与听写相似。平时多练习听写有助于记录效果的提高，而快速、准确的记录又有利于听力理解。所以，听与听写不是截然分开的。平时应多找些专业英语听力材料，认真进行听写训练。听写时一定要忠实于播讲人的发音，不要放过任何细微读音、词汇和语法疑点。一遍听不懂就反复听，这样才能学到更多的知识。对于听写出来的内容要反复推敲，精益求精，不要满足于能把所有的词句听写出来而对文章内容有大概的理解，而要从整体上进行检查，看有没有理解不了的地方，即使是细小的疑问也不要放过。只有平时进行严格训练，才能使自己在专业英语课堂上成为听力理解的强者。

4. 听与阅读并举

在听专业性英语授课或讲座的过程中，遇到自己不熟悉的词汇或表达方式时，人们往往会想"这种表达方式见得太少了"。所谓"见得太少"反映了平时对专业英语文献阅读较少的情况。所以,应多读多看,从而能对专业词汇和复杂的表达方式有所了解,不至于初次听到手忙脚乱。

5. 听与背并举

许多人在听过几次专业英语演讲或报告后会觉得其格式和句型比较规范。因此平时多背诵一些专业英语短文,就会较为熟悉其风格和常用句型,也就容易听懂新的内容。有时听了上一句话后就能预测到要说的下一句话。当然,口头专业演讲或报告毕竟不像书面语言那样死板,所以在听的过程中也要灵活处理。

6. 听与想并举

"想"是要求大家在听专业性英语授课或报告的过程中做一个有心人,善于开动脑筋,不断进行归纳和总结,使学到的知识系统化,使记忆深刻化。例如,每学到一个新的单词,都应该在拼写、发音和释义三个方面与自己已经掌握的单词进行比较,找出异同点,以便形成联系记忆。对于语法结构也是一样,如果理解不了或理解错了,就要找语法书来看,搞清楚正确的理解应该是什么。

口语技巧:突破语音关

英语口语的好坏,或者说能否达到用英语顺利进行口头交流的目的,其发音正确与否相当关键。正如语言教育研究专家 William Acton 所说"发音障碍和错误,轻则使人感觉不佳,影响学习的积极性和自信心,重则造成信息传递错误",在讲英语时,要特别注意英汉语音之间的三大不同之处。

1. 元音不同

汉语无长、短音之分,所以练习英语发音时容易出现长音不长、短音不短的现象,造成辨音与发音错误。另外,英语中的双元音与汉语中的复韵音相近,都是复合元音,但英语的复合元音动程长,而汉语的复韵母则短。例如:ai 与 $/ai/ \rightarrow /a \rightarrow i/$。所以,在讲英语时,要注意长元音发音饱满,短元音要尽量发音短促。

2. 辅音不同

英语的辅音,清浊对应分明。发清辅音时声带不振动,发浊辅音时声带一定要充分振动。而汉语中,许多辅音的特征主要是送气和不送气。所以在发与汉语相似的浊辅音时,一定要注意克服汉语的影响,声带一定要振动。此外,汉语中的音节多为开音节,而英语中闭音节是普遍现象。因此,不要把英语中连缀着的每一个辅音都念成一个音节,也不要在英语闭音节后面添上个 $/ə/$ 音。例如:bed 不要读成 $/bedə/$。

3. 咬舌音

英语有一对咬舌辅音,而汉语中没有。所以,较好的办法是对着镜子练习这对辅音,以便能自然熟练地发出这对咬舌音。

Reading Material

High Performance Liquid Chromatography (HPLC)

The applicability of chemiluminescence reactions as a means of detecting compounds in liquid chromatography (LC) is based to a large degree on post column reactions. A primer on liquid chromatography (and high performance LC) can be found here; however, a brief description follows. This describes, in the main, HPLC chromatographic systems.

Components of High Performance Liquid Chromatography

Liquid phase samples (mixtures) are injected into a LC column usually using a syringe and a specially devised injection valve. The sample is swept into the chromatographic column by the flowing mobile phase and chromatographic separation occurs as the mixture travels down the column. Normal HPLC detectors detect the elution of a compound from the end of the column based on some physical characteristic such as ultraviolet light absorption, ability to fluoresce, or the difference in index of refraction between the analyte and the mobile phase itself. The majority of HPLC systems work this way.

An example schematic of a HPLC system is shown below Figure 11-1.

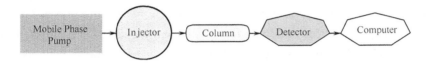

Figure 11-1 HPLC system

HPLC is a popular method of analysis because it is easy to learn and use and is not limited by the volatility or stability of the sample compound. The history section illustrates the HPLC's evolution from the 1970's to the 1990's. Modern HPLC has many applications including separation, identification, purification, and quantification of various compounds. It is important for those using HPLC to understand the theory of operation in order to receive the optimum analysis of their compounds. For those interested in purchasing or using a HPLC we have included a list of manufacturers, a troubleshooting guide, technical assistance, and a bibliography to help reduce your personal research and referencing time. Once you have completed the theory of operation, you will be qualified to take a quick quiz to test your understanding of HPLC systems.

Column Efficiency

Column efficiency refers to the performance of the stationary phase to accomplish particular separations. This entails how well the column is packed and its kinetic performance. The efficiency of a column can be measured by several methods, which may or may not be affected by chromatographic anomalies, such as "tailing" or appearance of a "front". This is important because many chromatographic peaks do not appear in the preferred shape of normal Gaussian distribution. For this reason efficiency can be an enigmatic value since manufacturers may use different methods in determining the efficiency of their columns.

Mobile Phase

The mobile phase in HPLC refers to the solvent being continuously applied to the column, or stationary phase. The mobile phase acts as a carrier for the sample solution. A sample solution is injected into the mobile phase of an assay through the injector port. As a sample solution flows through a column with the mobile phase, the components of that solution migrate according to the non-covalent interactions of the compound with the column. The chemical interactions of the mobile phase and sample, with the column, determine the degree of migration and separation of components contained in the sample. For example, those samples, which have stronger interactions with the mobile phase than with the stationary phase will elute from the column faster, and thus have a shorter retention time, while the reverse is also true. The mobile phase can be altered in order to manipulate the interactions of the sample and the stationary phase. There are several types of mobile phases, these include: isocratic, gradient, and polytypic.

Stationary Phase

The stationary phase in HPLC refers to the solid support contained within the column over which the mobile phase continuously flows. The sample solution is injected into the mobile phase of the assay through the injector port. As the sample solution flows with the mobile phase through the stationary phase, the components of that solution will migrate according to the non-covalent interactions of the compounds with the stationary phase. The chemical interactions of the stationary phase and the sample with the mobile phase, determines the degree of migration and separation of the components contained in the sample. For example, those samples, which have stronger interactions with the stationary phase than with the mobile phase will elute from the column less quickly, and thus have a longer retention time, while the reverse is also true. Columns containing various types of stationary phases are commercially available. Some of the more common stationary phases include: liquid-liquid, liquid-solid (adsorption), size exclusion, normal phase, reverse phase, ion exchange, and affinity.

Liquid-Solid operates on the basis of polarity. Compounds that possess functional groups cabable of strong hydrogen bonding will adhere more tightly to the stationary phase than less polar compounds. Thus, less polar compounds will elute from the column faster than compounds that are highly polar.

Liquid-Liquid operates on the same basis as liquid-solid. However, this technique is better suited for samples of medium polarity that are soluble in weakly polar to polar organic solvents. The separation of non-electrolytes is achieved by matching the polarities of the sample and the stationary phase and using a mobile phase which possesses a markedly different polarity.

Size-Exclusion operates on the basis of the molecular size of compounds being analyzed. The stationary phase consists of porous beads. The larger compounds will be excluded from the interior of the bead and thus will elute first. The smaller compounds will be allowed to enter the beads and will elute according to their ability to exit from the same sized pores they were internalized through. The column can be either silica or non-silica based. However, there are some size-exclusion that are weakly anionic and slightly hydrophobic which give rise to non-ideal size-exclusion behavior.

Normal Phase operates on the basis of hydrophilicity and lipophilicity by using a polar stationary phase and a less polar mobile phase. Thus hydrophobic compounds elute more quickly than do hydrophilic compounds.

Reverse Phase operates on the basis of hydrophilicity and lipophilicity. The stationary phase consists of silica based packings with n-alkyl chains covalently bound. For example, C-8 signifies an octyl chain and C-18 an octadecyl ligand in the matrix. The more hydrophobic the matrix on each ligand, the greater is the tendency of the column to retain hydrophobic moieties. Thus hydrophilic compounds elute more quickly than do hydrophobic compounds.

Ion-Exchange operates on the basis of selective exchange of ions in the sample with counterions in the stationary phase. IE is performed with columns containing charge-bearing functional groups attached to a polymer matrix. The functional ions are permanently bonded to the column and each has a counterion attached. The sample is retained by replacing the counterions of the stationary phase with its own ions. The sample is eluted from the column by changing the properties of the mobile phase do that the mobile phase will now displace the sample ions from the stationary phase, (changing the pH).

Affinity operates by using immobilized biochemicals that have a specific affinity to the compound of interest. Separation occurs as the mobile phase and sample passes over the stationary phase. The sample compound or compounds of interest are retained as the rest of the impurities and mobile phase pass through. The compounds are then eluted by changing the mobile phase conditions.

Injectors for HPLC

Samples are injected into the HPLC via an injection port. The injection port of a HPLC commonly consists of an injection valve and the sample loop. The sample is typically dissolved in the mobile phase before injection into the sample loop. The sample is then drawn into a syringe and injected into the loop via the injection valve. A rotation of the valve rotor closes the valve and opens the loop in order to inject the sample into the stream of the mobile phase. Loop volumes can range between 10 μl to over 500 μl. In modern HPLC systems, the sample injection is typically automated.

Stopped-flow injection is a method whereby the pump is turned off allowing the injecion port to attain atmospheric pressure. The syringe containing the sample is then injected into the valve in the usual manner, and the pump is turned on. For syringe type and reciprocation pumps, flow in the column can be brought to zero and rapidly resumed by diverting the mobile phase by means of a three-way valve placed in front of the injector. This method can be used up to very high pressures.

HPLC Pumps

There are several types of pumps available for use with HPLC analysis, they are: reciprocating piston pumps, syringe type pumps, and constant pressure pumps.

Reciprocating piston pumps consist of a small motor driven piston which moves rapidly back and forth in a hydraulic chamber that may vary from 35~400 μl in volume. On the back stroke, the separation column valve is closed, and the piston pulls in solvent from the mobile phase reservoir. On the forward stroke, the pump pushes solvent out to the column from the reservoir. A wide range of flow rates can be attained by altering the piston stroke volume during each cycle, or by altering the stroke frequency. Dual and triple head pumps consist of identical piston-chamber units which operate at 180 or 120 degrees out of phase. This type of pump system is significantly smoother because one pump is filling while the other is in the delivery cycle.

Syringe type pumps are most suitable for small bore columns because this pump delivers only a finite volume of mobile phase before it has to be refilled. These pumps have a volume between 250 ml and 500 ml. The pump operates by a motorized lead screw that delivers mobile phase to the column at a constant rate. The rate of solvent delivery is controlled by changing the voltage on the motor.

In constant pressure pumps the mobile phase is driven through the column with the use of pressure from a gas cylinder. A low-pressure gas source is needed to generate high liquid pressures. The valving arrangement allows the rapid refill of the solvent chamber whose capacity is about 70 ml. This provides continuous mobile phase flow rates.

Selected from "Knox, J. H. and Kauer, B.; High Performance Liquid Chromatography; Brown. P. R. and Hartwick, R. A. Eds.; Wiley Interscience: New York, 1989, Chapter 4."

Unit 12

Text: Mass Spectrometry

Introduction

Mass spectrometry, infrared spectroscopy, ultraviolet spectroscopy, and nuclear magnetic resonance spectroscopy are the four most common tests to determine the structure of an organic compound. These four tests answer the basic questions we have on size and formula, what functional groups are present, what is a conjugated pi electron system present, and what carbon-hydrogen framework is present. These are questions that in till the past few decades have required time-consuming research.

Mass Spectrometry

The technique for measuring the mass, and therefore the molecule mass of a mole is called Mass Spectrometry (MS). It is possible to also gain information about a molecule by measuring the fragments produces when molecules are broken apart. One of the most common kinds of mass spectrometers is electron-ionization (see Figure 12-1).

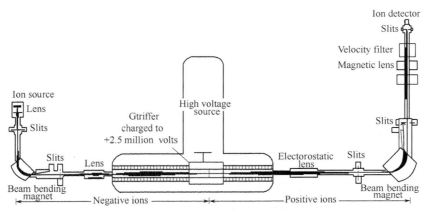

Figure 12-1 Structure of electron-ionization

The electron-ionization works by putting "A small amount of sample is vaporized in to the mass spectrometer, where it is bombarded by a stream of high-energy electrons. The energy of the electron beam can be varied but is commonly around 70 electron volts (eV). When a high-energy electron strikes an organic molecule, it dislodges a valence electron from the molecule, producing a cation radical. Electron bombardment transfers so much energy to the molecules that most of the cation radicals fragment after formation. The fragments flow through a curved pipe in a strong magnetic field, which deflects them by slightly different amounts according to their mass-to-charge ratio, neutral fragments are not detected by the magnetic field and are lost on the walls of the pipe, but positively charged fragments are sorted by the mass spectrometer onto a detector, which records them as peaks at various m/z ratios."

A mass spectrum of a compound is usually presented as bar graph with masses on the x-axis and intensity on the y-axis. Base peaks are the tallest peak and are assigned an intensity of 100% (see Figure 12-2).

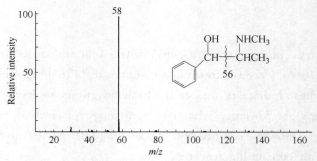

Figure 12-2　Electron ionization of ephedrine at 70 eV.

No molecular ion is seen (m/z 165)

The Mass Spectrometer

Mass spectrometry uses an instrument called mass spectrometer. The different components of the mass spectrometer are:

- Inlet system
- Ion source
- Mass analyzer
- Detector

Selected from "DiCesare, J.L.; Dong, M.W.; Vandermark, F.L.; Am.Lab., 1981, No.13, p.52."

New Words

mass spectrometry　质谱学，质谱分析
ultraviolet　*adj.* 紫外线的，紫外的；*n.* 紫外线辐射
nuclear magnetic resonance　核磁共振
conjugated　*adj.* 共轭的，成对的
fragment　*n.* 碎片，断片，片段
beam　*n.* 梁，桁条，（光线的）束，柱，电波，横梁
bombard　*vt.* 炮轰；轰击
stream　*n.* 流，一股，一串
dislodge　*v.* 驱逐
valence　*n.* [化]（化合）价，原子价
cation radical　阳离子基
electron bombardment　电子轰击

magnetic field 磁场
deflect v. （使）偏斜，（使）偏转
sort n. 种类，类别；v. 分类，拣选
bar graph 条线图，柱状图，线状光谱
axis n. 轴
intensity n. 强烈，剧烈，强度
component n. 成分；adj. 组成的，构成的
inlet n. 进口，入口，水湾，小港，插入物
ion source 离子源

Notes

1. Mass spectrometry, infrared spectroscopy, ultraviolet spectroscopy, and nuclear magnetic resonance spectroscopy are the four most common tests to determine the structure of an organic compound.

   参考译文：质谱、红外光谱、紫外光谱和核磁共振是四种最常见的测定有机化合物结构的方法。

2. A small amount of sample is vaporized into the mass spectrometer, where it is bombarded by stream of high-energy electrons.

   参考译文：少量的试样经过汽化后进入质谱仪，在那里试样受到高能电子流的轰击。

3. The fragments the flow through a curved pipe in a strong magnetic field, which deflects them by slightly different amounts according to their mass-to-charge ratio, neutral fragments are not deflected by the magnetic field and are lost on the walls of the pipe, but positively charged fragments are sorted by the mass spectrometer onto a detector, which records them as peaks at various m/z ratios.

   参考译文：在一个强大的磁场中分子碎片通过一个弯曲导管，由于质荷比不同发生的偏转程度略有不同。磁场不能使电中性的分子碎片偏转而被遗留在导管壁上。但那些带有正电荷的碎片却被质谱仪传送到检测器上，在检测器上记录不同的质荷比所达到的峰值。

4. A mass spectrum of a compound is usually presented as bar graph with masses on the x-axis and intensity on the y-axis. Base peaks are the tallest peak and are assigned an intensity of 100%.

   参考译文：我们通常用条线图来表示一种化合物的质谱图：x 轴表示质荷比，y 轴表示强度。基峰为最高的峰值并且强度值为 100%。

Exercises

1. Put the following into Chinese.

   ultraviolet spectroscopy   mass spectrometry
   nuclear magnetic resonance   structure of an organic compound
   carbon-hydrogen framework   electron-ionization
   mass-to-charge ratio   inlet system

2. Put the following into English.

红外光谱	官能团
价电子	阳离子基
检测系统	离子源

3. Questions.

(1) What are the four most common tests to determine the structure of an organic compound?

(2) How can you define the term "mass spectrometry" in simple English?

(3) In what way is a mass spectrum usually presented?

(4) How many different components of the mass spectrometer are there?

科 技 英 语 基 本 技 巧

听力技巧：多听相关专业的英语授课或报告

随着科学技术的发展，各种学科的知识和术语互相渗透的现象越来越多。近年来有了 computer virus（计算机病毒）的说法以后，vaccine, vaccination, immune, immunization 等医学术语相继渗入计算机的术语中。另一方面，同一个词在不同的学科领域或不同的专业中往往具有不同的词义。例如：名词 power 在不同的学科、专业中具有不同的含义：power source（电源），power unit（动力头，执行机构，功率单位），power shaft（传动轴），power operation（机械操作），power feed（自动进给）等，所以对于学习专业英语的人员来说应该广开"听路"。对于生僻的专业术语，可以通过上下文猜测或查词典识记。这不仅有助于扩大英语的知识面和词汇量，而且有助于克服一听到陌生的专业英语材料就害怕的恐惧心理。如此坚持一段时间后会发现，在听自己专业范围内的英语讲课或报告时，感觉会相对容易和轻松。

口语技巧：坚持"五说法"

练英语口语首先必须克服和消除腼腆和怕说错的心理障碍，并积极主动地创造条件，抓住一切机会去演练。其具体做法如下。

1. 勿腼腆大胆说

大胆说、不怕错才能使情绪保持镇定，保证发音器官正常工作、口语流畅，从而收到良好的训练效果。

2. 没有对象独自说

根据自己所处的地点、场合、环境，所见所闻的人、事、物，独自选定练口语主题，用英语表达，自言自语训练。

3. 遇到机会主动说

充分利用课余时间，参加一些"English corner"" English day "" English Bedroom",

这类的活动。这类活动具有一定的强制性,在规定的时间或地点内只许用英语交谈。在此类活动中,一定要主动搭话,养成用英语答话的习惯。

4. 进入环境积极说

只要遇上有外国人在场或举行外事活动的场合,就不应错过这种宝贵的机会,在有关部门的准许和不妨碍公务的情况下,只要有可能进入英语环境,就应积极争取,一旦进入英语环境,就应积极说、主动练。

5. 即景生情发挥说

根据不同的场合、地点、节日、活动等情景,借题发挥,用英语表达和描述所见情景时,扩展发挥的内容越多越好。

Reading Material

Nuclear Magnetic Resonance Spectroscopy

The nuclei of all elements carry a charge. When the spins of the protons and neutrons comprising these nuclei are not paired, the overall spin of the charged nucleus generates a magnetic dipole along the spin axis, and the intrinsic magnitude of this dipole is a fundamental nuclear property called the nuclear magnetic moment. The symmetry of the charge distribution in the nucleus is a function of its internal structure.

The Basis of NMR

The principle behind NMR is that many nuclei spin and all nuclei are electrically charged. In a magnetic field, spinning nuclei have lower energy when aligned with the field than when opposed to it because they behave like magnets. This energy difference corresponds to radio frequencies hence the nuclei are able to absorb and reemit radio waves. The diagram shows the case for the spin half nucleus. The principle is the same although more complex for higher spins

Uses of NMR spectroscopy

Nuclear Magnetic Resonance (NMR) spectroscopy is an analytical chemistry technique used for determining the content and purity of a sample as well as its molecular structure. For example, NMR can quantitatively analyze mixtures containing known compounds. For unknown compounds, NMR can either be used to match against spectral libraries or to infer the basic structure directly. Once the basic structure is known, NMR can be used to determine molecular conformation in solution as well as studying physical properties at the molecular level such as conformational exchange, phase changes, solubility, and diffusion. In order to achieve the desired results, a variety of NMR techniques are available.

All nuclei in molecules are surrounded by electrons. When an external magnetic field is

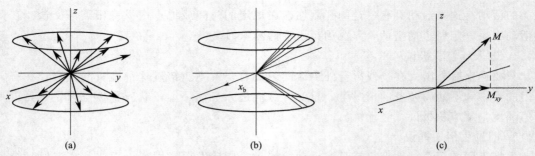

Figure 12-3 Molecules surrounded by magnetic field

applied to a molecule, the moving electrons set a tiny local magnetic fields of their own (see Figure 12-3). In describing this effect, we say that nuclei are shielded from the full effect of the applied field by the circulating electrons that surround them. If the NMR instrument is sensitive enough, the tiny differences in the effective magnetic fields experienced by different nuclei can be detected, and we can see a distinct NMR signal for each chemically distinct carbon, or hydrogen nucleus in a molecule. The NMR spectrum of an organic compound effectively maps the carbon-hydrogen framework.

Selected from "Novotny, M. Analytical Chemistry, 1989, Vol. 60, pp. 500~510."

PART 3 INSTRUMENTS

Unit 13

Text: Instruments

Buret

A buret is used to deliver solution in precisely measured, variable volumes. Burets are used primarily for titration, to deliver one reactant until the precise end point of the reaction is reached.

Using a Buret

To fill a buret, close the stopcock at the bottom and use a funnel. You may need to lift up on the funnel slightly, to allow the solution to flow in freely.

You can also fill a buret using a disposable transfer pipet. This works better than a funnel for the small, 10 ml burets. Be sure the transfer pipet is dry or conditioned with the titrant, so the concentration of solution will not be changed.

Before titrating, condition the buret with titrant solution and check that the buret is flowing freely. To condition a piece of glassware, rinse it so that all surfaces are coated with solution, then drain. Conditioning two or three times will insure that the concentration of titrant is not changed by a stray drop of water.

Check the tip of the buret for an air bubble. To remove an air bubble, whack the side of the buret tip while solution is flowing. If an air bubble is present during a titration, volume readings may be in error. Rinse the tip of the buret with water from a wash bottle and dry it carefully. After a minute, check for solution on the tip to see if your buret is leaking. The tip should be clean and dry before you take an initial volume reading.

When your buret is conditioned and filled, with no air bubbles or leaks, take an initial volume reading. A buret reading card with a black rectangle can help you to take a more accurate reading. Read the bottom of the meniscus. Be sure your eye is at the level of meniscus, not above or below. Reading from an angle, rather than straight on, results in a parallax error.

Deliver solution to the titration flask by turning the stopcock. The solution should be delivered quickly until a couple of ml from the end point.

The end point should be approached slowly, a drop at a time. Use a wash bottle to rinse the tip of the buret and the sides of the flask.

Pipet

A pipet is used to measure small amounts of solution very accurately. A pipet bulb is used to draw solution into the pipet.

Using a Pipet

Start by squeezing the bulb in your preferred hand. Then place the bulb on the flat end of the pipet. Place the tip of the pipet in the solution and release your grip on the bulb to pull solution into the pipet. Draw solution in above the mark on the neck of the pipet. If the volume of the pipet is larger than the volume of the pipet bulb, you may need to remove the bulb from the pipet and squeeze it and replace it on the pipet a second time, to fill the pipet volume completely.

Quickly, remove the pipet bulb and put your index finger on the end of the pipet. Gently release the seal made by your finger until the level of the solution meniscus exactly lines up with the mark on the pipet. Practice this with water until you are able to use the pipet and bulb consistently and accurately.

Volumetric Flask

A volumetric flask is used to make up a solution of fixed volume very accurately. This volumetric flask measures 500 ml $\pm$ 0.2 ml. This is a relative uncertainty of 4×10^{-4} or 400 parts per million.

Using Volumetric Flask

To make up a solution, first dissolve the solid material completely, in less water than required to fill the flask to the mark.

After the solid is completely dissolved, very carefully fill the flask to the 500 ml mark. Move your eyes to the level of the mark on the neck of the flask and line it up so that the circle around the neck looks like a line, not an ellipse. Then add distilled water a drop at a time until the bottom of the meniscus lines up exactly with the mark on the neck of the flask. Take care that no drops of liquid are in the neck of the flask above the mark. After the final dilution, remember to mix your solution thoroughly, by inverting the flask and shaking.

Titration

Begin by preparing your buret, as described on the buret page. Your buret should be conditioned and filled with titrant solution. You should check for air bubbles and leaks, before proceding with the titration.

Doing a Titration

Take an initial volume reading and record it in your notebook. Before beginning a titration, you should always calculate the expected end point volume.

Prepare the solution to be analyzed by placing it in a clean Erlenmeyer flask or beaker. If your sample is a solid, make sure it is completely dissolved. Put a magnetic stirrer in the flask and add an indicator.

Use the buret to deliver a stream of titrant to within a couple of ml of your expected end point. You will see the indicator change color when the titrant hits the solution in the flask, but the color change disappears upon stirring.

Approach the end point more slowly and watch the color of your flask carefully. Use a wash bottle to rinse the sides of the flask and the tip of the buret, to be sure all titrant is mixed in the flask (see Figure 13-1).

Figure 13-1 Doing titration

As you approach the end point, you may need to add a partial drop of titrant. You can do this with a rapid spin of a teflon stopcock or by partially opening the stopcock and rinsing the partial drop into the flask with a wash bottle.

Make sure you know what the end point should look like. For phenolphthalein, the end point is the first permanent pale pink. The pale pink fades in 10 to 20 minutes.

If you think you might have reached the end point, you can record the volume reading and add another partial drop. Sometimes it is easier to tell when you have gone past the end point. When you have reached the end point, read the final volume in the buret and record it in your notebook.

Subtract the initial volume to determine the amount of titrant delivered. Use this, the concentration of the titrant, and the stoichiometry of the titration reaction to calculate the number of moles of reactant in your analyte solution.

Selected from "Schoeff, M. S. and Williams, R. H. Principles of Laboratory Instruments, Mosby: St. Louis, 1993."

New Words

buret   *n*. 滴定管，玻璃量管
stopcock   *n*. [机] 管闩，活塞，活栓，旋塞阀
glassware   *n*. 玻璃器具类
rinse   *v* （用清水）刷，冲洗掉，漂净
drain   *vt*. 排出； *vi*. 排水，流干
tip   *n*. 顶，尖端

bubble   *n.* 泡沫
whack   *vt.* 重打，击败；*vi.* 重击
wash bottle   洗瓶
leak   *vi.* 漏，泄漏   *vt.* 使渗漏
rectangle   *n.* 长方形，矩形
meniscus   *n.* 新月，弯液面，凹凸透镜
parallax error   视差，判读误差
pipet   *n.* 吸量管，移液管
squeeze   *n.* 压榨，挤；*v.* 压榨，挤，挤榨
bulb   *n.* 球形物
index finger   食指
volumetric flask   （容）量瓶
ellipse   *n.* 椭圆，椭圆形
invert   *adj.* 转化的；*vt.* 使颠倒
Erlenmeyer flask   锥形（烧）瓶，爱伦美氏（烧）瓶
magnetic   *adj.* 磁的，有磁性的，有吸引力的
stirrer   *n.* 搅拌器，搅拌者，搅拌用勺子
spin   *v.* 旋转；*n.* 旋转
teflon   ［化］聚四氟乙烯（塑料，绝缘材料）

Notes

1. Conditioning two or three times will insure that the concentration of titrant is not changed by a stray drop of water.

   参考译文：将滴定管润洗2～3次可以确保滴定剂的浓度不会由于偶尔滴进一滴水而发生改变。

2. When your buret is conditioned and filled, with no air bubbles or leaks, take an initial volume reading. A buret reading card with a black rectangle can help you to take a more accurate reading.

   参考译文：当滴定管经过检验之后将其装满，观察没有气泡或漏水现象时，读取初始值。使用一个黑色的矩形的读数卡有助于读取得准确。

3. Draw solution in above the mark on the neck of the pipet. If the volume of the pipet is larger than the volume of the pipet bulb, you may need to remove the bulb from the pipet and squeeze it and replace it on the pipet a second time, to fill the pipet volume completely.

   参考译文：将溶液吸取至移液管管颈的标线之上。如果移液管的容积大于吸耳球的容积，你需要将吸耳球从移液管上拿下来，挤压吸耳球然后再一次重新吸取，以便使移液管充分装满溶液。

4. Move your eye to the level of the mark on the neck of the flask and line it up so that the circle around the neck looks like a line, not an ellipse. Then add distilled water a drop at a time until the bottom of the meniscus lines up exactly with the mark on the neck of the

flask。

参考译文：将视线移至容量瓶瓶口的刻度线处并与其保持水平，这样瓶颈处的一周看上去就像一条线而不是椭圆。然后滴加蒸馏水，一次一滴，直到弯液面的底部正好与容量瓶瓶颈处的刻度线吻合。

5. If you think you might have reached the end point, you can record the volume reading and add another partial drop. Sometimes it is easier to tell when you have gone past the end point. When you have reached the end point, read the final volume in the buret and record it in your notebook.

参考译文：如果认为有可能已经达到滴定终点，就将容积的读数记录下来，然后在滴加半滴。有时很容易就能判断出滴定终点是否已滴过。当达到滴定终点时，读取滴定管上的最后读数，然后把它记录在笔记本上。

Exercises

1. Put the following into Chinese.

    stopcock   a buret reading card
    parallax error   glassware
    parts per million   meniscus
    air bubble   precisely measure

2. Put the following into English.

    滴定管   洗瓶
    容量瓶   烧杯
    锥形瓶   吸耳球

3. Questions.

    (1) What is the main usage of a buret?
    (2) How can you fill the buret?
    (3) What do you need to do to condition the buret before the titrating?
    (4) Do you know the usage of a pipet bulb?
    (5) How to do a titration?

科技英语基本技巧
写作技巧：短文缩写阶段

训练指导的方针是训练效果好坏的一个前提条件。合理地设置训练程序，使英语写作从初级到高级沿着一条循序渐进、由简到繁的进程发展是成功训练者必须具备的指导思想。我们认为，在写作训练的初期，应采纳一条从有材料可依的写作方式过渡到脱离材料进行自由写作方式的途径。从有材可依到无材可依的训练过程应包括三个阶段。

短文缩写（Summary）可以是就所学课文进行缩写，也可以采用其他阅读材料，但要求被缩写的材料难易程度不超过所学课本。被用于进行缩写的课文或其他材料必须观点明确、层次分明、叙述有条理。缩写时应做到简明扼要，抓住重点，不要拖泥带水，没有主次。初学阶段的被缩写材料不宜太长，以不超一千词为佳，缩写文以不超过200词为佳。以下就一篇短文进行缩写，限于篇幅，短文内容有所节略。

　　Most shops in Britain open at 9：00 a.m. and close at 5：00 or 5：30 in the evening. Small shops usually close for an hour at lunchtime. On one or two days a week-usually Thursday and/or Friday-some large food shops stay until about 8.00 p：m. for late night shopping.

　　Many shops are closed in the afternoon on one day a week. The days are usually Wednesday or Thursday and it is a different day in different towns. Nearly all shops are closed on Sunday. Newspaper shops are open in the morning, and sell sweet sand cigarettes as well. But there are legal restrictions on selling many things on Sundays. Many large food shops （supermarkets） are self-service. When you go into one of these shops you take a basket and you put the things you wish to buy it. You queue up at the cash desk and pay for everything just before you leave.

　　If anyone tries to take things from a shop without paying they are almost certain to be caught. Most shops have store detectives who have the job of catching shoplifters. Shoplifting is considered a serious crime by the police and the courts.

　　When you are waiting to be served in a shop, it is important to wait your turn. It is important not to try to be served before people who arrived before you. Many people from overseas are astonished at the British habit of queuing.

　　将短文缩写如下：

　　This article tells us about British shops. British shops usually open at 9：00 a.m. and close at 5：00 or 5：30 p.m. Many shops are closed in the afternoon one-day a week. Nearly all shops are closed on Sundays. In Britain, many large food shops are self-service. And when you wait to be served in a shop, you have to wait patiently for your turn.

　　这是一篇不超过100词的缩写，句子基本上由原文各段落的主要内容构成。个别段落被完全删除以保证缩写重点突出，前后连贯。缩写是一种"依材剪贴"的写作方式，基本上采用原材料中的词语和句子，仅作了部分调整，是最初级的写作方式。

听力技巧：篇幅方面的策略

　　专业英语听力内容篇幅长短不一，有单句，也有简短的对话、专业性报道、通知，还有长篇报告和讲座。专业长篇和短篇内容听力理解的侧重点是不同的。

　　听短篇内容时，应在语音、连读等方面多加注意。同时应考虑动词的前后搭配。同样的动词，其后面的介词不同，意思上就会发生重大变化。另外还应注意各种不同的句式，如陈述句、倒装句或省略句等。对于长篇专业报告或讲座，听力的侧重点应转移到其他方面。首先，要把注意力放到强调理解和记忆信息内容方面，如主题、特点、细节、归纳总结等。听到什么重要细节，就在心中默念和重复，以便加深记忆。其次，应注意段落与段落之间意义的关联。长篇专业性讲座或报告的段落意义发生变化时往往比较迟缓，几个句子表达同样或近似意思的现象很多。即使漏听某一句，也不影响对段落的总体理解。因此，千万不要因为某一句话没听懂就轻易停止"接收"信息。

PART 3 INSTRUMENTS

Reading Material

Science Lab Techniques

Measuring Mass with an Electronic Balance

The electronic balance has many advantages over other types of balance. The most obvious is the ease with which a measurement is obtained. All that is needed is to place an object on the balance pan and the measurement can be read on the display to hundredths of a gram. A second advantage, using the Zero button on the front of the balance, is less recognized by students of science beginning. Because one must never place a chemical directly on the balance pan, some container must be used. Place the container on the balance and the mass of the container will be displayed. By pressing the Zero button at this point, the balance will reset to zero and ignore the mass of the container. You may now place the substance to be weighed into the container and the balance will show only the mass of the substance. This saves calculation time and effort. However, when the container is removed from the balance, the display will go into negative numbers until the Zero button is pressed again.

Our electronic balances also have a Unit button on the front. Pressing this button will change the units being measured. Since we have very few times when we need something other than metric units, you should not have to change the mode on the balance. Because of the Unit and Zero buttons, there are two things you must always do before placing objects onto an electric balance to be measured:

- See that the display is reading 0.00
- See that the unit sign in the upper right of the display shows g

When finished with the electronic balance, press the ON/OFF button and hold it down until the display shows OFF.

Filtering a Precipitate from a Solution

Filtering a solid out of a liquid is done using filter paper and a filter funnel. The filter funnel is supported by the ring on a ring stand. Lay a clay triangle across the ring, and then place the filter funnel into the triangle.

To prepare the filter paper, fold the paper in half, then fold it in half again. When you look at the open edge of the folded paper you will see four edges of paper. With thumb and finger, catch three of these edges. Squeeze the sides of the folded paper and a cone will form with three thicknesses of paper on one side and one thickness of paper on the other. Place this cone of paper into the filter funnel. Place a "catch container" under the stem of the filter funnel and adjust the height of the ring on the ring stand until the tip of the stem is below the mouth of the container.

Use a wash bottle and wet down the inside of the filter paper. This will help it stick to the funnel. You are now ready to filter. Carefully pour the liquid to be filtered into the mouth of the funnel. Do not let the liquid rise to the top of the filter paper. If any liquid goes over and around the paper, your procedure is ruined. Be patient, it will take time for the liquid to move through the pores of the paper. When all the original liquid has been poured into the funnel, use the wash bottle to rinse any remaining precipitate out of the original container. Do not touch or try to stir the liquid inside the filter paper. The wet paper is easily torn, which will ruin your procedure.

If the objective of your filtration is the solid-free liquid, throw the filter paper and its contents into the trash. If your objective is the solid, carefully remove the filter paper and set it in a secure place to dry.

Using Litmus Paper to Determine Acid or Base

An acid turns blue litmus paper red and a base turns red litmus paper blue. While even grade-school students know this, there is one mistake commonly made when using either litmus or pH paper strips. You should never dip the test paper into the solution being tested. While the degree to which this contaminates the solution is not great, good chemistry students know not to do it. Always use a glass stirring rod. Dip a clean stirring rod into the solution, then touch the wet stirring rod to the paper.

Using a Magnetic Stirrer

A magnetic stirrer is helpful for dissolving solids in liquids. While there are several different styles, all will at least have a base with a speed-controlled spinning magnet inside and an external stirring bar. The stirring bar is placed into a flask or beaker by gently sliding it along the wall of the container.

To prevent breakage, do not drop the bar onto the bottom of the container. Place the container on the stirrer base and turn the speed control knob to its lowest setting. Use just enough speed to start the bar turning in the container. The picture here shows the "vortex" that forms inside the continer. Be patient, the bar might "hop" at excess speeds, causing splashing.

If using exact volumes, as with a volumetric flask, be sure take the measurements before adding the stirring bar.

When stirring is completed, keep the stirring bar in the container by "decanting" the liquid into another container. Be sure to carefully wash the original container and the stirring bar.

Distillation

In general chemistry, distillation can be used to remove dissolved substances from a liquid or to separate a mixture of liquids that have different boiling points.

To successfully separate a mixture of liquids, you must know the boiling point of each liquid involved and be able to measure temperature changes as heat is applied. Heat is added to the mixture until it reaches the first boiling point. The temperature must be kept at this temperature until all the first liquid is removed. The temperature is then allowed to rise to the next boiling point and the second liquid is removed. The closer these boiling points are together, the more difficult it is to collect pure liquids from the mixture.

One distillation apparatus is pictured at right. No matter how the apparatus is set up, the following things must be accomplished.
- The original liquid is heated
- The temperature is measured
- The vapor is collected and condensed back into a liquid
- The new liquid is collected

Generating and Collecting Gases

A gas generating bottle is used to produce gases from a chemical reaction. The solid is placed into the bottle, then a solution is poured down the funnel. As the gas builds, it will move into the collecting tube at the top of the bottle. If the gas were lighter than air, it would not go down into the container. Gases that are lighter than air must be collected by water displacement.

For water displacement, the container is filled with water then inverted into a water trough. The gas is bubbled up under the container and pushes the water out. This is a very common procedure in general chemistry labs and has only one draw-back. If the gas being generated is highly soluble in water, precise measurements of the gas produced will not be possible.

Fume Hood

The fume hood is a safety glass-front cabinet with an exhaust fan. It is used for experiments known to produce noxious fumes or smoke. Do the following to perform an experiment in the hood.
- Raise the door of the hood
- Turn on the light and set up the apparatus
- When all material for the experiment is ready, turn on the fan
- Pull the hood door at least 1/3 way down
- Perform the experiment
- When finished, pull the door all the way down until all smoke and fumes are removed
- Turn off the fan and the light, then remove the equipment to a regular lab station for cleaning
- Leave the hood clean

Hotwater Bath

A hotwater bath is used in the general chemistry lab to heat something slowly and evenly. Many different types of glassware can be used to set up a hotwater bath. What glassware you use will depend on the quantity of the substance you wish to heat. All hotwater baths put the substance to be heated in a container and then place that container into a container of water. Heat is then applied to the water container. The substance is heated by the water, not the burner flame.

In addition to heating slowly and evenly, water baths are also safer than direct heat. For this reason, highly volatile substances should only be heated using hotwater baths.

Selected from "Schoeff, M. S. and Williams, R. H. Principles of Laboratory Instruments, Mosby: St. Louis, 1993."

PART 3 INSTRUMENTS

Unit 14

Text: Spectrometer

The scanning spectrometer measures absorbance and wavelength for liquid samples. The spectrometer is controlled by a computer and the results are printed on an inkjet printer.

Using the Scanning Spectrometer

- The sample compartment is in the middle of the spectrometer and opens by lifting the cover.
- Sample cuvettes are shaped like this. They have two opposite sides that are cloudy and two opposite sides that are transparent, for the light to pass through. Place your cuvette in the sample compartment, with the transparent sides aligned with the light path (see Figure 14-1).

Figure 14-1 Place cuvette

- You will measure the spectrum of a blank before you use the spectrometer and the software will automatically subract the blank absorbance at each wavelength. To measure your spectrum, you will use the computer mouse to select Start, from the buttons along the bottom of the screen.
- When your spectrum is finished, you can print it on the printer.

Digital Spectrometer

A digital spectrometer measures the amount of visible light absorbed by a colored solution. This can be read as Absorbance or % Transmittance.

Setting Up the Spectrometer

- Next press the A/T/C button to select Absorbance mode. The current mode appears on the display.

- Set the wavelength to the desired value using the nm buttons. The wavelength is displayed on the LCD.
- Fill a cuvette with your blank solvent and dry the outside of the cuvette carefully.
- Insert your blank cuvette in the sample compartment and close the cover.
- Press 0 ABS/100% T to set the absorbance of the blank to zero.
- After a few moments the absorbance and wavelength will be displayed. The absorbance should be zero.
- Remove the blank and insert a dry sample cuvette.
- The absorbance value of the sample will now be shown on the LCD.
- You are now ready to calibrate the colorimeter. Prepare a blank by filling a cuvette 3/4 full with distilled water. To correctly use a colorimeter cuvette, remember:
- All cuvettes should be wiped clean and dry on the outside with a tissue.
- Handle cuvettes only by the top edge of the ribbed sides.
- All solutions should be free of bubbles.
- Always position the cuvette with its reference mark facing toward the white reference mark at the right of the cuvette slot on the colorimeter.

Place the blank cuvette in the cuvette slot of the colorimeter and close the lid. Turn the wavelength knob of the colorimeter to the 0%T position. In this position, the light source is turned off, so no light is received by the photocell. Type "0" in the % edit box. When the displayed voltage reading for Input 1 stabilizes, turn the wavelength knob of the colorimeter to the red LED position (635 nm). In this position, the colorimeter is calibrated to show 100% of the red light being transmitted through the blank cuvette. Type "100" in the % edit box.

At this point, you have prepared the colourimeter to properly measure transmittance and hence, absorbance, from any sample you wish, at a wavelength of 635 nm. Any measurements at other wavelengths will incure some errors unless you repeat the calibration procedure for those wavelengths. You are now ready to collect absorbance data for the most concentrated of the standard solutions. Empty the water from the cuvette. Using the most concentrated solution, rinse the cuvette twice with 1 ml amounts and then fill it 3/4 full. Wipe the outside with a tissue and place it in the colorimeter. After closing the lid, wait for the absorbance value displayed on the monitor to stabilize. Then type "635" in the edit box, and press the ENTER key. The data pair you just collected should now be plotted on the graph. Repeat this collection at the other two wavelengths (green and blue) without recalibrating.

Once you have all three points collected, you can determine the best wavelength λ_{max} to do the Beer's.

Selected from "Willard, H.; Merritt, L.; Dean, J.A.; Settle Jr., F.A. Instrumental Methods of Analysis, 7th ed., Wadsworth Publishing Co., 1988."

New Words

absorbance　*n*.［植、化］吸光度
cuvette　*n*. 比色皿，透明小容器，试管
transparent　*adj*. 透明的，显然的，明晰的
align　*vi*. 排列；*vt*. 使结盟，使成一行
blank　*n*. 空白
digital　*adj*. 数字的，数位的，手指的；*n*. 数字，数字式
desired value　期望值
button　*n*. 纽扣，［计］按钮
LCD (liquid crystal display)　*n*. 液晶显示器
insert　*vt*. 插入，嵌入
calibrate　*v*. 校准，使标准化，标定
colorimeter　*n*. 色度计，色量计
wipe　*v*. 擦，揩，擦去
tissue　*n*. 薄的纱织品，薄纸，棉纸
ribbed　*adj*. 有肋骨的，有棱纹的
free of　无……的，在……外面，摆脱……的
reference mark　［测］基准标记
white reference　基准白（色）
slot　*n*. 缝，狭槽
lid　*n*. 盖子
photocell　*n*. 光电池

Notes

1. You will measure the spectrum of a blank before you use the spectrometer and the software will automatically subtract the blank absorbance at each wavelength.
   参考译文：在使用分光光度计测量前，先测量出空白值，软件会自动地扣除在每一波长下的吸光度空白值。

2. A digital spectrometer measures the amount of visible light absorbed by a colored solution. This can be read as absorbance or % transmittance.
   参考译文：数字式分光光度计用来测定有色溶液吸收的可见光的数量。有色溶液中吸收的可见光的数量可以通过读取吸光度或透光率值得到。

3. Any measurements at other wavelengths will incur some errors unless you repeat the calibration procedure for those wavelengths.
   参考译文：测量其他波长时，应重复进行校准的过程，否则测量就会出现一些误差。

Exercises

1. Put the following into Chinese.
   liquid sample　　　　　　　　　　　　sample cuvette

colored solution　　　　　　　　　distilled water
visible light　　　　　　　　　　reference mark

2. Put the following into English.

吸光度　　　　　　　　　　　　空白
基准标记　　　　　　　　　　　期望值
色度计　　　　　　　　　　　　分光光度计
波长　　　　　　　　　　　　　透光率

3. Questions.

(1) What are sample cuvettes shaped like?

(2) How can you place a cuvette in the sample compartment?

(3) What is the usage of a digital spectrometer?

(4) How can we set up the spectrometer?

(5) There are many things to remember before you use a colorimeter cuvette correctly. Can you point them out?

Lab Key

科 技 英 语 基 本 技 巧
科技英语写作句型：问题

句型 1. 这是研究 ×× （与 ×× 有关）的问题

　　A. this is a problem

　　B. dealing with （concerned with）
　　　 bearing on （relating to）
　　　 which deals with （which bears on）

　　C. the nature of （the influence of, the role of）…
　　　 the measurement of （the activity of）…

句型 2. 某问题容易（难于）解决（详细讨论）

　　A. the problem of （under investigation, you mention here）
　　　 the problem, as has been outlined, （as we have just seen）

　　B. is easy （not too complicated）
　　　 seems difficult
　　　 does not seem too intricate （far too involved, to broad）

　　C. (to solve very quickly)
　　　 (to formulate in precise terms)
　　　 (to present in all its complexity)

句型 3. 某问题在 ×× 范围之内（外）

　　A. this problem you have just outlined （I have referred to, authors have raised）

B. is within (falls)
      lies beyond (goes)
      seems to be outside
   C. the scope of the above theory
      the range of the existing theory
      the province of the problems pertaining to …
      the limits of the given research (this theory, our concepts, our present knowledge of …)

句型 4. 要解决某问题对××需进一步研究
   A. (much) further research
      a great deal of further research effort (investigation)
      further work
   B. is needed to elucidate
      is required to discover (clarify)
      is wanted to bring to light (reveal)
   C. certain aspects of (details of) …
      some facts about …
      the intricate nature of …

句型 5. 某问题××角度未考虑
   A. the problem of …
   B. is interpreted (was regarded)
      has been viewed
      can be looked upon (be interpreted)
   C. in terms of concepts
      in the light of this fundamental theory (recent findings in this area)
      as a new one (a fundamental one in the field)
      in the main one

听力技巧：题材方面的策略

专业英语听力材料比较丰富，其中主要包括叙述性、说明性和议论性材料。这方面材料各有其特点，因此听到专业性报告或讲座时应"对症下药"。

1. 叙述性材料

它主要在谈话、对话中出现，其特征在于叙述"问题"与"反应"的关系。一旦遇到这种材料应把注意力放在其核心问题上，即谈的是什么问题，讲话的人对此反应如何。

2. 说明性材料

它是通过举例、定义、分类、对比等手法来说明、解释某一事物，使其较清楚地展现在听众面前。在专业性讲座和学术讨论会上经常使用这种手法。它具有一定的结构模式，即下定义—分类—举例说明—结论。因此应把听力重点放在对主题句的理解上。

3. 议论性材料

这种材料在学术讨论会或科技交流会上经常出现。它在论述某件事情或某种事实时往往提出两种或多种观点，有时还有对立的观点存在。因此在听力的过程中，要特别注意论点与反论点，抓住其关键的对立部分。

Reading Material

Single-Beam Spectrometer

The Spectronic 20 is an example of a direct-reading, single-beam spectrophotometer. By virtue of being direct-reading, they feature fast operation, and because of their simple design they are relatively inexpensive and require a minimum of maintenance. They are particularly well suited for spectrophotometric determinations of a single component at a single wave length where only moderate accuracy ($\pm 1\% \ T$ to $\pm 3\% \ T$) is required.

The limitations of the single-beam instrument become readily apparent when an absorption spectrum over a wavelength range is required. The response of the phototube, the emissivity of light source, and the intensity of the light diffracted by the grating are all a function of wavelength (see Figure 14-2). Accordingly, in order to obtain the absorption spectrum of a compound, the instrument must be recalibrated each time the wavelength setting is changed.

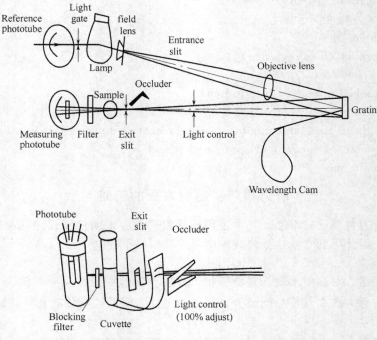

Figure 14-2 Structure of single-beam spectrometer

Experiment

First, adjust the instrument, photocell dark, with the zero control. Next, insert into the sample holder a test tube filled with water. Rotate the 100% control until the meter needle reads near mid scale. Now rotate the wavelength control with a maximum transmittance reading is achieved. Readjustment of the 100% control may be necessary to keep the meter needle on scale.

Carefully balance the instrument at 100% transmittance at the wavelength of maximum transmittance. Without changing either the dark current control or the 100% T control, vary the wavelength from 350 nm to 625 nm in 25 nm intervals and record the instrument response. Plot the instrument response to wavelength.

The phototube in the Spectronic 20 is type S-4 (cesium-antimony). The relative response of the phototube to a beam of monochromater light of constant intensity is greatest near 400 nm and decreases fairly rapidly above 475 nm. Its response at 625 nm is only 5% of that at its maximum near 400 nm.

Handling of Cuvettes

The handling of the cuvette is extremely important. Often two cuvettes are used simultaneously, one for the "blank" solution and one for the sample to be measured. Yet any variation in the cuvette (such as a change in cuvette width or the curvature of the glass, stains, smudges or scratches) will cause varying results. Thus it is essential, in dealing with cuvettes, to follow several invariant rules.

(1) Do not handle the lower portion of a cuvette (through which the light beam will pass).

(2) Always rinse the cuvette with several portions of the solution before taking a measurement.

(3) Wipe off any liquid drops or smudges on the lower half of the cuvette with a clean Scott wiper (or other lens paper) before placing the cuvette in the instrument. Never wipe cuvettes with towels or handkerchiefs. Inspect to ensure that no lint remains on the outside and that no small air bubbles are clinging to the inside walls.

(4) When inserting a cuvette into the sample holder:

a. To avoid any possible scratching of the cuvette in the optical path, insert the cuvette with the index line facing toward the front of the instrument;

b. After the cuvette is seated, line up the index lines exactly.

(5) When using two cuvettes simultaneously, use none of the cuvettes always for the blank solution and the other cuvette for the various samples being measured. Mark the tubes accordingly and do not interchange the cuvettes during the remainder of the course.

Selected from "Willard, H.; Merritt, L.; Dean, J.A.; Settle Jr., F.A. Instrumental Methods of Analysis, 7th ed., Wadsworth Publishing Co., 1988."

Unit 15

Text: Clarus 400 GC Guide

Introduction

The Clarus 400 Gas Chromatograph is a dual-channel, temperature-programmable stand-alone gas chromatograph (GC). It is available in many configurations, such as with or without an autosampler and a variety of injector/detector combinations to provide you with total GC flexibility. The Clams 400 GC is microprocessor controlled, where you enter the operating parameters from the color-coded keyboard and view the prompting text and monitor instrument functions on a large two-line vacuum fluorescence display.

Clarus 400 GC may have none, one, or two of the following detectors installed:

Flame Ionization Detector (FID)　　Nitrogen Phosphorus Detector (NPD)
Electron Capture Detector (ECD)　　Thermal Conductivity Detector (TCD)

The FID, ECD, TCD, or the NPD, may be installed in either the front or the rear detector position. Each installed detector has one analog output which may be attached to either an integrator or recorder. Signals may be routed under instrument control.

About the Keyboard

The keyboard is your link to the software. The keyboard has 35 keys divided into the following groups:

- Function keys
- Parameter keys
- Entry keys
- Control keys

Note: As you run this instrument you will see software functions on the display that are not supported by the Clarus 400 GC. Please ignore these functions and continue with your analysis. An audible short beep sounds every time a key is pressed. A long beep sounds when an error has been made.

Installing a Packed Column

The packed column injector consists of a septum cap, needle guide, quartz injector liner, and the injector body. This injector is used with 1/8 inch or 1/4 inch glass or metal packed columns. In addition, by installing the 530 micron wide-bore Adapter Kit, you can convert the injector to accept wide-bore capillary columns.

Step 1: Turn off the Heaters

The moment the Clarus 400 GC is turned on, the oven, injector (s), and detector (s)

PART 3　INSTRUMENTS

begin to heat up rapidly. To avoid burns and injury while installing a column, all heaters should be turned off and their respective zones allowed to cool before touching the injector septum caps or any of the fittings inside the oven.

It is recommended that you remove the injector liner shipped with the packed injector and pack it with a small amount of silanized glass wool before performing analyses.

Step 2: Set the Carrier Gas Flow

1. Turn on the carrier gas at the tank.
2. Adjust the line pressure to 620 kPa.
3. Press [**Carrier Prog**] until the appropriate screen appears.

```
Flow 1   30
Set    : 30ml · min⁻¹
```
(Flow 1   30 / Set : $30\text{ml} \cdot \text{min}^{-1}$)

4. Type the desired flow setpoint value and press [**Enter**].
5. Adjust the flow by turning the flow control knob counterclockwise to increase the flow, clockwise to decrease the flow, until the actual flow displayed equals the set point value.
6. Attach a soap bubble flowmeter to the packed injector fitting.
7. Turn on the carrier gas at the tank and adjust the line pressure to 90 psi (1psi = 6894.76Pa).
8. Press [**System**] [**System**] [**Enter**] to display the stopwatch screen.
9. Start the carrier gas flowing by turning the flow controller knob counterclockwise.
10. Measure the flow.

NOTE: For best accuracy, use a soap bubble flowmeter volume or electronic flowmeter that gives a reading of at least 30 seconds.

11. Adjust the flow to the desired set point by repeatedly measuring the flow and turning the flow controller knob counterclockwise to increase the flow, clockwise to decrease the flow, until the desired flow is obtained.
12. Disconnect the soap bubble flowmeter before proceeding to the next step.

Step 3: Connect One End of the Column to the Packed Injector

NOTE: If you are installing a 1/4 inch column, attach a 1/8 inch to 1/4 inch adapter to the packed injector fitting before continuing. Finger tighten the adapter, then while holding the packed injector fitting steady with a 7/16 inch wrench, tighten the adapter with a 9/16 inch wrench.

1. Insert one end of the column into the packed injector fitting until it bottoms, then finger tighten the column nut onto the packed injector fitting.
2. While holding the packed injector fitting with one 7/16 inch wrench, tighten the column nut an additional 1/8 to 1/4 turn with the other wrench. Do not over tighten column nuts. Over tightening causes permanent damage to the fittings.

Step 4: Leak Test

Test the connection to the packed injector fitting for leaks using a 50/50 mixture of isopropanol/water or an electronic leak detector. To avoid contaminating the system, do not use a soap solution for leak testing. Tighten all leaking connections.

Step 5: Condition the Column

This section contains a suggested temperature program for conditioning a column. The program starts off by holding the oven temperature at a medium value for 10 minutes, gradually increasing the oven temperature at a fixed rate ($5\,^{\circ}\!C \cdot min^{-1}$) to the column operating temperature, then holding that temperature overnight with the carrier gas flowing. The temperatures shown in the following examples should only be used as guidelines. Please refer to the column manufacturer's operating instructions for specific temperature recommendations.

1. Close the oven door, then press [**Oven Prog**]. The **Oven Temperature** screen appears.

Oven NOT RDY	30°
TEMP 1	: 75°

2. Enter an oven temperature set point of 50, then press [**Enter**]. The **Oven Time** screen appears:

Oven NOT RDY	0.0m
TIME 1	: 999.9m

3. Enter a (Hold) TIME of 10, then press [**Enter**]. The **Oven Rate** screen appears:

Oven NOT RDY	30°
RATE 1	: End

4. To add another program step, enter a RATE of 5 ($^{\circ}\!C \cdot min^{-1}$). A screen similar to the following appears:

Oven NOT RDY	40°
TEMP 2	: 50°

5. For TEMP 2, enter a set point 25℃ to 50℃ above your planned analytical operating temperature. For example, enter a set point of 150.

Oven NOT RDY	50°
TEMP 2	: 150°

CAUTION: To avoid damaging the column, do not enter a temperature higher than the maximum operating temperature specified by the column manufacturer.

6. Press [**Enter**]. The next screen is:

Oven	0.0m
TIME 2	: 999.9m

7. Press [**Enter**]. The next screen is:

Oven	NOT RDY
RATE 2	: End

8. Set an Injection Temperature about 50℃ higher than the TEMP 2 setting.

9. Turn Detector Temperature off. Press [**RUN**] and allow the system to run overnight.

10. The next morning press [**Reset Oven**]. A menu similar to the following appears:

```
Reset to Oven Temp
: 1 2
```

11. Press [**Enter**]. This resets the oven temperature set point to that specified for TEMP 1 at the beginning of the temperature program.

12. Open the oven door, then press [**CE**].

Allow the oven to cool until the oven fan goes off. This occurs when the oven cools down to 40℃.

NOTE: Condition a new column before using it in an analysis. Once it is conditioned, you will not need to recondition it.

Step 6: Attach the Other End of the Column to the Detector

1. Insert the free end of the column into the detector fitting, then finger tighten the column nut onto the detector fitting.

2. While holding the detector fitting with one of the 7/16 inch wrenches, tighten the column nut an additional 1/8 to 1/4 turn with the other wrench. Make certain that no part of the column touches the bottom or sides of the oven once it is installed.

NOTE: If you are installing a 1/4 inch column, attach a 1/8 inch to 1/4 inch adapter to the detector fitting before continuing. Finger tighten the adapter, then while holding the detector fitting steady with a 7/16 inch wrench, tighten the adapter with a 9/16 inch wrench.

Step 7: Leak Test the Column/Detector Connection

The following procedures describe leak testing the column to detector connections. With the carrier gas still flowing from the overnight conditioning, test the column/detector connection for leaks using a 50/50 mixture of isopropanol/water or use an electronic leak detector. To prevent contaminating the system, do not use a soap solution for leak testing. Tighten the connection if a leak is found. Set up the detector to be used with this column.

Selected from "PerkinElmer Life And Analytical Sciences, Inc. 940 Winter Street Waltham, Massachusetts 02451 USA (781) 663-6900"

New Words

dual *adj.* 双的，二重的，双重的
programmable *adj.* 可设计的，可编程的
stand-alone *n.* 单机

chromatograph n. 色谱仪
configuration n. 构造，结构，配置，外形
autosampler n. 自动进样器
injector n. 注射器
microprocessor n. 微处理器
parameter n. 参数，参量
fluorescence n. 荧光，荧光性
Flame Ionization Detector（FID）（氢）火焰离子化检测器
Nitrogen Phosphorus Detector（NPD）氮磷检测器
Electron Capture Detector（ECD）电子捕获检测器
Thermal Conductivity Detector（TCD）热导检测器
integrator n. 积分仪
keyboard n. 键盘
audible adj. 听得见的
beep n. 哔哔声；v. 嘟嘟响
septum n. 隔膜
quartz n. 石英
liner n. 衬垫
adapter n. 适配器，改编者
injury n. 伤害，损坏
silanize vt. 使硅烷化
wool n. 羊毛，毛织品，毛线，绒线，毛料衣物
counterclockwise adj. 逆时针方向的；adv. 逆时针方向地
clockwise adj. 顺时针方向的；adv. 顺时针方向地
soap n. 肥皂
flowmeter n. 流量计
disconnect v. 拆开，分离，断开
wrench n. 扳钳，扳手
nut n. 坚果，螺母，螺帽
isopropanol n. 异丙醇
leak n. 漏出，漏出物，泄漏；vi. 漏，泄漏；vt. 使渗漏
manufacturer n. 制造者，厂商

Notes

1. The Clarus 400 Gas Chromatograph is a dual-channel, temperature-programmable stand-alone gas chromatograph (GC). It is available in many configurations, such as with or without an autosampler and a variety of injector/detector combinations to provide you with total GC flexibility.

   参考译文：Clams 400 型气相色谱仪是一种双通道的、可程序控制升温的单机气相色谱

仪。该型号的气相色谱仪有许多种配置，例如有带一个自动进样器，也有不带自动进样器，而且还有各种各样的注射器/检测器组合可以提供 GC 系列最大限度的灵活性。

2. The Clarus 400 GC is microprocessor controlled, where you enter the operating parameters from the color-coded keyboard and view the prompting text and monitor instrument functions on a large two-line vacuum fluorescence display.

参考译文：Clarus 400 型气相色谱仪是由微处理器控制的。在带有色码的键盘上输入操作所需要的操作参数后，在一个比普通型号大一倍的真空荧光显示屏上即可以看到提示文本，按照提示文本就可以监控仪器的功能。

3. The FID, ECD, TCD, or the NPD, may be installed in either the front or the rear detector position. Each installed detector has one analog out put which may be attached to either an integrator or recorder.

参考译文：（氢）火焰离子化检测器、电子捕获检测器、热导检测器或氮磷检测器既可以安装在检测器的前部，也可以安装在检测器的后部。每个检测器安装好后都有一个模拟信号输出到积分仪或记录仪上。

4. The packed column injector consists of a septum cap, needle guide, quartz injector liner, and the injector body. This injector is used with 1/8 inch or 1/4 inch glass or metal packed columns. In addition, by installing the 530 micron wide-bore Adapter Kit, you can convert the injector to accept wide-bore capillary columns.

参考译文：填充柱注射器由隔膜帽、针头、石英注射器衬垫和注射器壳体组成。注射器与 1/8in（英寸）或 1/4in 的玻璃填充柱或金属填充柱配套使用。另外，通过安装 530μm 的大孔适配器，可以改装注射器与大孔毛细管柱配套使用。

5. The program starts off by holding the oven temperature at a medium value for 10 minutes, gradually increasing the oven temperature at a fixed rate (5℃·min^{-1}) to the column operating temperature, then holding that temperature overnight with the carrier gas flowing.

参考译文：程序开始时先将炉温保持中等温度 10min，然后逐渐地按照固定的速率（5℃·min^{-1}）将炉温提高到填充柱所需要的操作温度，接着通入载气将操作温度保持一夜的时间。

Exercises

1. Put the following into Chinese.

gas chromatograph	injector
packed column	flowmeter
Flame Ionization Detector	integrator

2. Put the following into English.

色谱柱	检测器
载气	自动进样器
异丙醇	记录仪

3. Questions.

   (1) How many detector configurations can be installed in the Clarus 400 GC?

   (2) How many keys are there in the keyboard? How many groups can these keys be di-

vided into? And what are they?

(3) What does the packed column injector consists of?

(4) How many steps are involved in installing the packed column? And what are they?

(5) How to set the carrier gas flow according to this passage?

科 技 英 语 基 本 技 巧

写作技巧：科技论文英文摘要的类型

摘要的定义为：以提供文献内容梗概为目的，不加评论和补充解释，简明、确切地记叙文献重要内容的短文。由于大多数检索系统只收录论文的摘要部分，或其数据库中只有摘要部分免费提供，并且有些读者只阅读摘要而不读全文或常根据摘要来判断是否需要阅读全文，因此摘要的清楚表达十分重要。好的英文摘要对于增加论文的被检索和引用机会、吸引读者、扩大影响起着不可忽视的作用。

根据内容的不同，摘要可分为以下三大类：报道性摘要、指示性摘要和报道-指示性摘要。

1. 报道性摘要

报道性摘要（informative abstract）也称信息型摘要或资料性摘要。一般包括了原始文献某些重要内容的梗概，主要由以下三部分组成。

（1）目的：主要说明作者写此文章的目的，或说本文主要解决的问题。

（2）过程及方法：主要说明作者工作过程及所用的方法，也包括众多的边界条件、使用的主要设备和仪器。

（3）结果：作者在此工作过程中最后得到的结果和结论，如有可能，尽量提一句作者所得结果和结论的应用范围和应用情况。

2. 指示性摘要

指示性摘要（indicative abstract）也称为说明性摘要、描述性摘要（descriptive abstract）或论点摘要（topic abstract）。一般只用二、三句话概括论文的主题，而不涉及论据和结论，多用于综述、会议报告等。此类摘要可用于帮助读者决定是否需要阅读全文。

3. 报道-指示性摘要

报道-指示性摘要（informative-indicative abstract）以报道性摘要的形式表述一次文献中信息价值较高的部分，以指示性摘要的形式表述其余部分。传统的摘要多为一段式，在内容上大致包括引言（Introduction）、材料与方法（Materials and Methods）、结果（Results）和讨论（Discussion）等主要方面，即 IMRAD（Introduction, Methods, Results and Discussion）结构的写作模式。

20世纪80年代出现了另一种摘要文体，即"结构式摘要"（structured abstract），它是报道性摘要的结构化表达，强调论文摘要应含有较多的信息量。结构式摘要与传统摘要的差别在于，前者便于读者了解论文的内容，行文中用醒目的字体（黑体、全部大写或斜体等）直接标出目的、方法、结果和结论等标题。

Reading Material

Clarus 400 GC Specification

The PerkinElmer® Clarus® 400 Gas Chromatograph (GC) provides the proven dependable performance and analytical capabilities of the Clarus GC family. It is available with manual pneumatic control, single channel or dual-channel configurations and optional integral liquid autosampler for diverse application needs.

Oven

The Clarus 400 GC oven provides easy access to columns. The oven gives excellent temperature control for maximum productivity.

Volume:	10600cm^3	
Temperature range:	10℃ above ambient to 450℃	
Column overheat protect:	User settable up to 450℃	
Temperature program:	3 ramp, 4 plateaus	
	Minimun Range	**Increment**
Oven temperature:	10℃ above ambient to 450℃	1℃
Initial time:	0 to 999 min	0.1min
Rate:	0.1 to 45℃·min^{-1}	0.1℃
Plateau time:	0 to 999 min	0.1min
Cool-down times:	250℃ to 50℃	4.8min
	200℃ to 50℃	3.8min

Carrier gas pneumatics

- Carrier gas pneumatics are included with the Clarus 400 GC injectors.
- Manual pneumatics are available for all injectors.
- Two carrier zones.
- Split-vent pneumatics are included with the Clarus 400 GC split/splitless capillary injector.

Detector pneumatics

Manual pneumatics are available for all detectors.

Autosampler

The Clarus 400 GC offers an optional, built-in syringe autosampler for maximum sampling capabilities. All control is accomplished through the keypad or by a data system.

Injection speed:	Normal, fast, slow
Program mode:	Two methods may be programmed
Number of sample positions:	82, plus one priority
Vial size:	2ml(0.25ml with insert)crimp-top caps
	2ml screw-top caps
Number of waste and wash vials:	4 waste and 4 wash
Waste and wash vial size:	4ml
Syringe size:	0.5μl, 5.0μl or 50.0ul

续表

Sampling volume:	0.1μl to 0.5μl from the 0.5μl syringe in 0.1μl increments
	0.5μl to 5.0μl from the 5.0μl syringe in 0.5μl increments
	5.0μl to 50.0μl from the 50.0μl syringe in 5.0μl increments
Viscosity settings:	0-15
Max number of injections/vial:	15
Max number of solvent postwashes:	15
Max number of sample pumps:	15
Max number of sample prewashes:	15
Min sample volume required:	5ul when used with the 0.25ml vial insert; 350μl when used with the 2ml vial
Reproducibility:	<0.5% RSD for packed columns 1% C_9 in C_7. 1μl injected

Injectors

The Clarus 400 GC supports split/splitless capillary and packed column injectors that provide accuracy and precision to all of your sampling applications. Up to two injectors may be installed and operated simultaneously with independent temperature control. Every injector is available with manual pneumatics.

Split/splitless capillary injector

• Split ratio easily adjustable for a wide range of analysis conditions.
• Charcoal trap in split vent prevents contamination of split valve and lab air.
• Two choices of liner: 2mm and 4mm internal diameter.
• 50℃ to 450℃ in 1℃ increments.
• 1/16 inch fitting.
• Manual pneumatics—pressure regulator (0-60 psi) for digital display of column—head pressure.
• Automatic control of split vent solenoid valve.

Packed column injector

• Removable glass liner for trapping nonvolatile residues.
• Adapter for on-column injection to wide-bore capillary columns.
• 50℃ to 450℃ in 1℃ increments.
• 1/8 inch fitting.
• 1/4 inch column adapter available.
• Manual pneumatics—choice of flow controller with head-pressure gauge, or flow controller with headpressure gauge and digital display of flow.

Detectors

A choice of four detectors, optimized for sensitivity and selectivity, is available for use with the Clarus 400 GC. All built-in detectors include an automated background compensation feature that corrects for column bleed. Whether you choose the Flame Ionization Detector, the Thermal Conductivity Detector, the Electron Capture Detector or the Nitrogen Phosphorus Detector, all conform to the highest industry standards for reliability and performance.

Flame Ionization Detector (FID)
- Wide linear dynamic range.
- No makeup gas required due to efficient sweeping of column effluent by hydrogen combustion gas.
- Air flow designed to minimize contamination and residue buildup.
- 1/8 inch fittings.
- Manual pneumatics—pressure regulator for hydrogen, needle valve for air.
- "Flame out" warning and ready interlock.

Operating temperature:	100℃ to 450℃ in 1℃ increments
Sensitivity:	>0.015 C/(g·C)
Min detectable quantity:	$<3\times10^{-12}$(g·C)/s(nonane at a $S/N=2$ to 1)
Linearity:	$>10^6$
Signal filtration:	50ms, 200ms, 800ms
Input range:	1, 20
Makeup gas:	Not required

Electron Capture Detector (ECD)
- High sensitivity.
- Excellent selectivity.
- High operating temperature for maximum stability.
- 1/8 inch fittings.
- Manual pneumatics-needle valve for makeup gas.

Source:	15mCi^{63}Ni
Temperature protect:	470℃ by software
Carrier gas:	Either Ar/CH$_4$ or N$_2$
Operating temperature:	100℃ to 450℃ in 1℃ increments
Min detectable quantity:	<0.05 pg perchloroethylene with argon/methane or nitrogen
Linearity:	$>10^4$
Signal filtration:	200ms, 800ms
Makeup gas:	Standard

Thermal Conductivity Detector (TCD)
- Capillary-column compatible.
- Proven constant current design.
- Software protection to prevent filament burnout.

- Ideal for series operation.
- 1/8inch fittings.
- Manual pneumatics-reference gas flow controller.

Operating temperature:	100℃ to 350℃ in 1℃ increments
Sensitivity:	$9\mu V \cdot ppm^{-1}$ (nonane at 160 mA at the bridge with a detector temperature of 100℃)
Min detectable quantity:	Typically<1 ppm(nonane, 1ppm=10^{-6})
Linearity:	>10^5
Power supply:	Constant current with four selectable settings: 1: ±40mA 2: ±80mA 3: ±120mA 4: ±160mA
Signal filtration:	50ms, 200ms, 800ms
Filament protection:	Self-limiting and resetting after transient overloads in either channel
Makeup gas:	Not required for 0.32mm to 0.53mm i.d. columns with flows>$5ml \cdot min^{-1}$
	Required for 0.25mm or smaller i.d. columns

Nitrogen Phosphorus Detector (NPD)

- Modular design.
- Change bead in less than one minute.
- Pre-aligned bead.
- Rapid conditioning—up and running in less than two hours.
- 1/8inch fittings.
- Manual pneumatics—pressure regulator for hydrogen, needle valve for air.

Operating temperature:	100℃ to 450℃ in 1℃ increments
Min detectable quantity:	5×10^{-13} (g·N)/s(2,4-dimethylaniline)
	5×10^{-14} (g·P)/s(tributylphosphate)
Linearity:	>10^4
Signal filtration:	50ms, 200ms, 800ms
Selectivity:	50,000 : 1(N/C)
	10 : 1(P/N)
Input range:	1, 20
Makeup gas:	Not required

Physical details

Power requirements:	120V±10%, 50/60Hz±1%, 2.4kV·A*
	220V±10%, 50/60Hz±1%, 2.2kV·A
	230V±10%, 50/60Hz±1%, 2.3kV·A
	* On an independent 20amp line
Ambient temperature:	10℃ to 32℃
Ambient humidity:	20% to 80% relative humidity without condensation
Altitude:	Operating: sea level to 2000m
	Non-operating: sea level to 12000m

PART 3 INSTRUMENTS

续表

Mean BTU output:	3400
Weight:	Clarus 400 GC without Autosampler:
	49kg(108 lb)
	Clarus 400 GC with Autosampler:
	54kg(118 lb)
Dimensions:	Clarus 400 GC without Autosampler:
	(H×W×D): 54cm×67cm×72cm(21in×26in×28in)
	Clarus 400 GC with Autosampler:
	80cm×67cm×72cm(32in×26in×28in)

Selected from "perkinElmer Life And Analytical Sciences. Inc. 940 Winter Street Waltham, Massachusetts 02451 USA (781) 663-6900"

Unit 16

Text: Optimizing PerkinElmer Flame AA

This short tutorial is designed to show the operator of a PerkinElmer flame atomic absorption spectrometer how to optimize the burner system. This material is primarily designed for new users, but can be used as a review and provide some additional hints on how to properly aligned the burner system. Please follow this presentation fully before adjusting your instrument.

Outline

The following items will be covered as part of this presentation. This material is designed for all manually adjusted models. The units with automatic alignment do not apply, however the sections on nebulizer are applicable. Determining when this procedure has to be performed visually checking the burner position. The procedure of aligning the burner position includes vertical, horizontal and rotational adjustments. Checking the nebulizer by preadjusting and measuring the uptake rate, adjusting the nebulizer for best sensitivity. Verifying that the adjustments have been done correctly.

When to Perform

Optimization should be performed after any of the following items have taken place. If you know the system is out of alignment. This may occur when the burner has been swapped out, the adjustments have been performed incorrectly or with a new installation. Low sensitivity has been observed with all elements. Be aware that if only one analyte is out, there could be something unique to this analyte that is not related to the burner alignment or nebulizer adjustment. Always compare values to previously measured or known values. Burner alignment and nebulizer adjustment are generally necessary after maintenance.

Preconditions

In order to perform any of these adjustments, the instrument must be properly set up. Choose an element, such as copper, that is easy to perform. Avoid using an element that is flame sensitive, such as chromium or gold. Make sure the lamp is on. Recall or create a method for this element. If you are using a stored method, make sure the method does not have background correction turned on. Place the software in Continuous Graphics or, in the case of older non-computer controlled instruments, the continuous mode.

- Make sure the unit is ready to perform adjustments.

- Properly copper or a lamp that emits visible (orange) light
- Lamp is on and aligned
- Create or open proper method
- Open Continuous Graphics to be in the continuous mode

Checking Burner Position

Before performing this task, make sure the flame is or has been shutdown. The quickest way to verify that the burner is near or close to proper alignment is by inserting a white card with a predetermined crosshair mark on a white card. The light of the hollow cathode lamp should project an image on the card. If you cannot see an image, verify that a light hollow cathode is in place. Lamp that project orange light are much more visible.

- Use a white card
- Using a previously drawn, aligned crosshair will help to be more accurate

Aligning the Burner (vertical)

Ensure that the Burner Head is below the optical beam. Select the Auto Zero Graph button. Adjust the vertical adjustment knob on the burner assembly to raise the burner head into the path of the beam until the absorbance reading begins to increase above zero. Lower the Burner Head until the absorbance reads zero and then adjust an additional 1/2 turn in the same direction.

- Lower the burner until light is well above burner head
- Auto zero
- Raise burner until reading increases
- Lower to zero and 1/2 turn lower

Aligning the Burner (Horizontal)

If not already done, light the flame and aspirate the sensitivity check standard. Adjust the burner position by turning the horizontal alignment control alternately while observing the continuous graphics display. Adjustment until you find the position that gives maximum absorbance reading. Repeat the horizontal adjustment, if you do not see any sensitivity or the sensitivity is very low, perform a nebulizer adjustment or verify that the solution is reaching the flame. The next slides will demonstrate the adjustment of the nebulizer.

- Light flame with blank.
- Auto zero
- Aspirate standard
- Adjust horizontal for highest reading
- If not reading, the nebulizer must be checked

Nebulizer Preadjustment and Checking

Aspirate the standard solution. This solution should be made to give a reading of approximately 0.2 absorbance. Refer to the Recommended conditions for more details. Turn the nebulizer adjustment knob slowly counter-clockwise until bubbles start to appear from the end of the sample tube in the standard solution. In the next slides we will adjust the nebulizer for the best readings.

- Light or verify that the flame is light
- Aspirate a known standard solution that should read approximately 0.2 absorbance
- Turn the nebulizer counter clockwise until bubbles appear at the capillary tubing

After the bubble have appeared, turn the nebulizer adjustment knob 1 to 1.5 turns clockwise. You should hear an audible change in the flame sound. Alternatively you can check to make sure the solution is getting into the flame by aspirating a solution of $2 \text{ mg} \cdot \text{L}^{-1}$ of sodium. The flame should turn yellow when the solution reaches the flame. Note: if using an autosampler, the time for the solution to reach the burner will be longer. You may disconnect the autosampler and use a shorter piece of tubing for this adjustment.

- Turn the nebulizer clockwise 1 to 1.5 turns.
- You should hear the change in the sound of the flame
- Alternatively: Na test
- $2 \text{ mg} \cdot \text{L}^{-1}$ Na will show a yellow flame

Note: If using an autosampler, the time for the solution to reach the burner will be longer.

Adjusting the Nebulizer

Slowly continue to turn the nebulizer adjustment knob clockwise while watching the continuous graphics display. The absorbance reading will increase to maximum before starting to decrease. Adjust the nebulizer to gained the maximum reading. You may see two or three peaks and valleys in the process of turning the nebulizer clockwise. Pick the highest of the peaks. By measuring the uptake rate and noting this value, you can use this as a reference for future checks, as a rule, the value of this is related to the readings achieved. If you do not see a change in sensitivity, and you know the solution is reaching the flame, the burner may be very far out of alignment or there may be other problems that are beyond the scope of this course.

- Aspirate the standard
- Adjust the nebulizer for maximum sensitivity
- Using a graduated cylinder measure the uptake rate
- This value can be used to check aspiration rates later

Verifying Proper Sensitivity

You should expect to get a predetermined value, This value is either determined by the recommended conditions or previously known data. Keep in mind that the sensitivity can be affected by it's presence. Refer to your instrument manual for more details.

Verify that the signal read is comparable to what is known or expected:
- "Recommended Conditions"
- Previous values

The sensitivity can be different and may be based on the following:
- Nebulizer type
- Impact device
- Length of tubing
- Spacer present (if part of nebulizer)

Selected from "PerkinElmer Life and Analytical Sciences 710 Bridgeport Avenue Shelton. CT 06484-4794 USA Phone: (800) 762-4000 or (+1) 203-925-4602 www.perkinelmer.com"

New Words

optimize   vt. 使最优化
PerkinElmer   （美国）珀金埃尔默仪器有限公司
flame   n. 火焰，光辉，光芒
tutorial   n. 指南
hint   n. 暗示，提示，线索
align   vt. 矫正（路线），（直线）对准
manually   adv. 用手，手动
nebulizer   n. 喷雾器
vertical   adj. 垂直的，直立的；n. 垂直线，垂直面，竖向
horizontal   adj. 地平线的，水平的
rotational   adj. 转动的，轮流的
verify   vt. 检验，校验，查证，核实
sensitivity   n. 敏感，灵敏（度），灵敏性
previously   adv. 先前，以前
maintenance   n. 维护，保持
precondition   v. 预处理
software   n. 软件
image   n. 图像，肖像，偶像，映像
hollow cathode lamp   空心阴极灯
optical   adj. 光学的
beam   n. 梁，桁条，（光线的）束，柱，电波，横梁
aspirate   v. 吸气

capillary　*n*. 毛细管；*adj*. 毛状的，毛细作用的
recommend　*vt*. 推荐，介绍，劝告

Notes

1. This short tutorial is designed to show the operator of a PerkinElmer flame atomic absorption spectrometer how to optimize the burner system.
   参考译文：该操作指南旨在向珀金埃尔默仪器有限公司出品的火焰原子吸收分光仪的操作用户展示如何优化燃烧器系统。
2. Turn the nebulizer adjustment knob slowly counter-clockwise until bubbles start to appear from the end of the sample tube in the standard solution.
   参考译文：按逆时针方向缓慢地旋转喷雾器的调整按钮，直到浸入到标准溶液里的管子末端出现气泡为止。
3. Adjust the vertical adjustment knob on the burner assembly to raise the burner head into the path of the beam until the absorbance reading begins to increase above zero.
   参考译文：为了使燃烧器的顶部提升至光路，不断调整燃烧器组件上的垂直调节按钮，直到吸光度读数开始增至正数为止。
4. Slowly continue to turn the nebulizer adjustment knob clockwise while watching the continuous graphics display. The absorbance reading will increase to maximum before starting to decrease.
   参考译文：按顺时针方向连续缓慢地旋转喷雾器的调整按钮，同时密切注意连续显示的图形。吸光度的读数会增至最大值后开始下降。

Exercises

1. Put the following into Chinese.
   hollow cathode lamp　　　　　　　　atomic absorption spectrometer
   optical beam　　　　　　　　　　　　nebulizer
   capillary tubing　　　　　　　　　　　autosampler
2. Put the following into English.
   灵敏度　　　　　　　　　　　　　　　火焰
   吸光度　　　　　　　　　　　　　　　水平调节
   仪器　　　　　　　　　　　　　　　　预处理
3. Questions.
   (1) When should optimization be performed on the burner system?
   (2) What is the quickest way to verify that the burner is near or close to proper alignment?
   (3) How can you check the burner position?
   (4) How can you align the burner vertically?
   (5) How many steps are involved in the adjustments the nebulizer?

科 技 英 语 基 本 技 巧
写作技巧：科技论文英文摘要的内容

为了国际交流，科学技术报告、学位论文和学术论文应附有外文（多用英文）摘要。科技论文英文摘要的内容包含题名、摘要及关键词。

一、英文题名

1. 题名的结构

英文题名以短语为主要形式，尤以名词短语最常见，即题名基本上由一个或几个名词加上其前置或后置定语构成。短语型题名要确定好中心词，再进行前后修饰。各个词的顺序很重要，词序不当，会导致表达不准。例如：

Thermodynamic Characteristics of Water Absorption of Heat-treated Wood.
热处理木材的水分吸着热力学特性

题名一般不应是陈述句，因为题名主要起标示作用，而陈述句容易使题名具有判断式的语义；况且陈述句不够精练和醒目，重点也不易突出。少数情况（评述性、综述性和驳斥性）下可以用疑问句做题名，因为疑问句可有探讨性语气，易引起读者兴趣。例如：

Can Agricultural Mechanization be Realized Without Petroleum?
农业机械化能离开石油吗？

2. 题名的字数

题名不应过长。国外科技期刊一般对题名字数有所限制。例如，美国医学会规定题名不超过2行，每行不超过42个印刷符号和空格；美国国立癌症研究所杂志 J. Nat Cancer Inst 要求题名不超过14个词；英国数学会要求题名不超过12个词。这些规定可供我们参考。总的原则是，题名应确切、简练、醒目，在能准确反映论文特定内容的前提下，题名词数越少越好。

3. 中英文题名的一致性

同一篇论文，其英文题名与中文题名内容上应一致，但不等于说词语要一一对应。在许多情况下，个别非实质性的词可以省略或变动。例如：

工业湿蒸汽的直接热量计算
The Direct Measurement of Heat Transmitted Wet Steam

英文题名的直译中译文是"由湿蒸汽所传热量的直接计量"，与中文题名相比较，二者用词虽有差别，但内容上是一致的。

4. 题名中的冠词

在早年，科技论文题名中的冠词用得较多，近些年有简化的趋势，凡可用可不用的冠词均可不用。例如：

The Effect of Ground Water Quality on the Wheat Yield and Quality

其中两处的冠词 the 均可不用。

5. 题名中的大小写

题名字母的大小写有以下 3 种格式。

（1）全部字母大写。

例如：OPTIMAL DISPOSITION OF ROLLER CHAIN DRIVE

（2）每个词的首字母大写，但 3 个或 4 个字母以下的冠词、连词、介词全部小写。

例如：The Deformation and Strength of Concrete Dams with Defects

（3）题名第 1 个词的首字母大写，其余字母均小写。

例如：Topographic inversion of interval velocities

目前（2）格式用得最多，而（3）格式的使用有增多的趋势。

6. 题名中的缩略词语

已得到整个科技界或本行业科技人员公认的缩略词语，才可用于题名中，否则不要轻易使用。

二、作者与作者单位的英译

1. 作者

中国人名按汉语拼音拼写，其他非英语国家人名按作者自己提供的罗马字母拼法拼写。

2. 单位

单位名称要写全（由小到大），并附地址和邮政编码，确保联系方便。前段时间一些单位机构英译纷纷采取缩写，外人不知所云，结果造成混乱。FAO，WHO，MIT 尽人皆知，而 BFU 是 Beijing Forestry University，恐怕只有圈内人才知道。另外，单位英译一定要采用本单位统一的译法（即本单位标准译法），切不可另起炉灶。

三、英文摘要

摘要一般不超过 250 单词，不少于 100 单词，少数情况下可以例外，视原文文献而定。但主题概念不得遗漏，另外，写、译或校文摘可不受原文文摘的约束。文摘叙述要简明，逻辑性要强，句子结构严谨完整，尽量用短句子，一般缩短文摘方法如下。

（1）取消不必要的字句：如"It is reported…""Extensive investigations show that…" "the author discusses…""This paper concerned with…"

（2）对物理单位及一些通用词可以适当进行简化；

（3）取消或减少背景情况（background information）；

（4）限制文摘只表示新情况、新内容，过去的研究细节可以取消；

（5）不说废话，如"本文所谈的有关研究工作是对过去老工艺的一个极大的改进"切不可进入文摘；

（6）作者在文献中谈及的未来计划不纳入文摘；

（7）尽量简化一些措辞和重复的单元。

四、关键词

关键词是英语科技论文的文献检索标识，用以表达文献主题概念的自然语言词汇。论文的关键词应由文题和层次标题选出。国际重要检索系统均用关键词来检索当前最新发表的论文。如果论文作者选用关键词不当，所用关键词不能正确表达论文原创性内容，可能会导致文献检索数据库无法识别而拒绝加以收录。

Reading Material

Features and Operation of Hollow Cathode Lamps

The Hollow Cathode Lamp

The primary requirement of a hollow cathode lamp is to generate a narrow emission line of the element which is being measured. This line should be of sufficient spectral purity and intensity to achieve a linear calibration graph with low noise level from the AA spectrometer.

When a discharge occurs between two electrodes via a gas at low pressure, the cathode is bombarded by the energetic, positively-charged gas ions (i. e. ionized filler gas atoms) which are accelerated towards its surface by the potential existing in the discharge. The energy of these ions is such that atoms of the cathode material are ejected or "sputtered" into the plasma. Here they may collide with other high energy particles, which are present. These collisions result in a transfer of energy causing the metal atoms to become excited. Since this excited state is not stable, the atoms relax back to their ground state, emitting radiation at the characteristic wavelength of that element. For most elements, more than one analytically useful spectral line is generated. In order to obtain optimum performance from the lamp, there are a number of design parameters which must be carefully selected.

Hollow Cathode Lamp Operation

There are two parameters of major importance which affect the analytical results. These are:
(1) The hollow cathode lamp current, which affects the intensity of the source, and
(2) The spectral band width (or slit width) of the instrument which affects the isolation of the spectral line.

In order to simplify the selection of these two parameters, Varian supplies information on the recommended operating conditions for each lamp. However in special cases improved results improved results may be achieved by minor deviations from the recommended conditions. The choice of operating conditions depends greatly on whether the operator is searching for maximum precision at concentrations near the detection limit or whether samples are being measured over a wide concentration range.

1. Lamp current

The principle effect of an increase in the lamp current is an increase in the intensity of the lamp emission. The intensity of the lamp affects the baseline (absorbance) noise level upon which the analytical signal is measured. This baseline noise level significantly affects the analytical performance in terms of precision and detection limits.

This noise level is inversely related to intensity of the light source and therefore the more intense the source, the lower the baseline noise will be.

If this were the only consideration then the lamp would be run at the maximum allowable current. However in practice it is not as simple as this.

As the operating current increases above the recommended value, increased broadening of the emission lines occurs until at very high currents the phenomenon of self-absorption is observed. This appears as an inversion of the peak top of the emission line due to a cloud of atoms in front of the cathode absorbing the cathodic emission within the lamp itself.

The resultant distortion of the lamp emission peak leads to reduced sensitivity.

This will also be passed on as more pronounced curvature in the calibration plot as shown for cadmium. It should be noted that this example is one of the most current-sensitive of elements. Other elements may show little or no effect from current variation.

Excessive lamp currents applied to the lamp will accelerate the sputtering process and will shorten the lifetime of the lamp. This is especially true for the more volatile elements.

For measurements near the detection limit (where baseline noise is important) there may be advantages in using a higher lamp current than is recommended. This is especially true for those elements which show little loss in absorbance with increased lamp current.

On the other hand, a lower lamp current may yield a more linear calibration for an extended range but at the expense of extra noise level.

It becomes apparent that a compromise may be required to obtain the best sensitivity coupled with a high signal-to-noise ratio and a long lamp life. Varian provides a set of recommended operating conditions with each lamp.

2. Lamp intensity

Each analytical line from each hollow cathode lamp has a characteristic intensity which relates to the observable signal-to-noise level of the atomic absorption instrument. The greater the intensity of the analytical line, the lower the noise level. Such differences in the measured noise level between different lamps are quite normal. For example the silver line at 328.1 nm has a greater intensity than the iron line at 248.3 nm and the resultant baseline noise levels illustrate this in the signal graphics traces.

It should be noted that the photocathode response characteristics of the photomultiplier tube will affect the noise level observed. Varian use a photomultiplier tube which has a high response over a wide wavelength range.

3. Spectral band width

The spectral band width (SBW) affects the spectral isolation of the analytical line. The spectral band width required is normally dictated by the nearest adjacent line in the spectrum.

The spectral scan of antimony can be seen that when using the most intense line at 217.6nm, a SBW of less than 0.3 nm is required to avoid interference from the 217.9 nm line. By studying the effect of altering the SBW on the absorbance of an analyte solution the optimum SBW can be determined.

4. Warm-up time

Stability of the hollow cathode lamp signal is very important. Typically hollow cathode lamps require a warm-up period after switch on, during which time the lamp achieves an equilibrium state and the output stabilizes.

Warm-up time is particularly important for single beam instrument operation. With single beam instruments (such as the SpectrAA-10) the change in intensity of the lamp is reflected in the baseline of the instrument, That is, the baseline drifts as the lamp drifts. Therefore it is important that the lamp be allowed sufficient time to warm-up prior to performing any analytical measurements. For the majority of elements 10 minutes is a suitable warm-up period to achieve a stable signal. Exceptions to this are As, P, Tl and the Cu/Zn multielement lamp where longer warm-up times are recommended.

With double beam optical configuration, the instrument is able to compensate for changes in the sample beam intensity by making continual comparison with the reference beam. This comparison is carried out at 50 or 60 Hz with the Varian SpectrAA series, giving a sample and reference beam measurement every 20 or 16 milliseconds respectively.

With double beam instruments, the lamp warm-up time is not apparent. Nevertheless it is desirable to allow a short Warm-up before attempting precise analytical measurements. This is because the profile of the emission line from the lamp can change during this period. and small changes in analytical signal may result.

With double beam instruments, the zero absorbance level will always be maintained.

Note also that the SpectrAA instrument offers true double beam operation when the analytical measurement occurs, although it is optically a single beam instrument.

5. Multielement lamps

Multielement lamps may consist of up to six different elements. The elements are combined in the cathode as metallurgical powders. Such lamps are convenient to the analyst, but there are some limitations to this approach.

Some combinations of elements cannot be used because their emission lines are so close together that they interfere with each other. The recommended conditions are often quite different from those for the single element lamp and should be strictly observed for best results.

Single element lamps yield superior quality results in terms of calibration linearity than for the same element in a multielement combination. Multielement lamps do, however, provide convenience of operation for less demanding analyses.

Selected from "K. Brodie and S. Neate, Varian Australia Pty. Ltd Melbourne, Victoria, Australia"

Unit 17

Text: Operation of the Mettler S20 pH meter

Basic Instrumentation Theory

The basic phenomenon upon which the pH meter operates is the development of an electrical potential (voltage) by a chemical reaction in an electrochemical cell. Such a cell consists of two dissimilar half cells, different in materials or concentration, separated by a semipermeable membrane. This allows for the contact of the two solutions but does not allow mixing and hence direct reaction of the substances in the two half cells. The voltage is measured across the two electrodes in the half cells.

The system consists of a pH sensor, pH Half Cell, whose voltage varies proportionately to the hydrogen ion activity of the solution, and a reference electrode, Reference Half Cell, which provides a stable and constant reference voltage.

The pH electrode consists of a thin membrane of hydrogen sensitive glass blown on the end of an inert glass tube. Because this is a special type of glass and very thin, the bulb is very fragile and great care must be exercised in handling it.

This tube is filled with an electrolyte, and the signal is carried through Ag/AgCl wire. This is a pH half Cell. A similar system, but without using a hydrogen sensitive glass, is used as a reference. A small filter connects this tube to the external liquid. This system is called a Reference Half Cell.

It is very common to find pH meters which are fitted with only a single combination electrode. As the name implies, a combination electrode is a combination of the glass electrode and the reference electrode into a single probe. The probe is constructed with the reference surrounding the glass electrode. The primary advantage to using a combination electrode is the ability to measure the pH of a smaller volume of sample or a sample in a container with a restricted opening. The main disadvantages to using a combination electrode are the more limited selection of internal elements (most have only a silver-silver chloride internal reference) and the higher cost. The combination electrode on your pH meter may have a plastic shield around the glass bulb to protect it from damage.

Operations of a pH Meter

Now let's turn our attention to the pH meter that you will be using in the laboratory. For this you must be at a work station with the following materials:
1. Handout
2. pH meter and electrodes
3. Solution Bottle A pH=7 Standardization Solution
4. Wash bottle (distilled water)

5. Beakers A and Waste (empty)
6. Tissue

As you work through this part of the module, you must perform certain operations with the pH meter. Do these operations carefully and only after you fully understand the operation so that the equipment will not be damaged.

Initial Set Up

1. Take the plastic cap off the bottom of the electrode. Save the solution (3mol·L^{-1} KCl) in the cap to store the electrode when finished. Do NOT discard the liquid inside the plastic cap but retain for storage.

2. Plug the meter in, if needed, and press the ON/OFF button to turn the meter on. The meter will do a quick self-check and the display will then show the pH, temperature and a number of other parameters.

3. Check that pH is the mode, shown by the pH on the upper right of the main display box. If not, use the pH/mV button to toggle to pH.

Calibrate the Meter

1. Rinse the electrode with distilled water and pat dry with a tissue.

Rinse the electrodes with distilled water from a wash bottle into an empty beaker.

Blot the electrodes dry with a tissue. Do not rub!

This procedure should be carried out whenever the electrodes are transferred form one solution to another to minimize the chance of contamination.

2. Place the electrode in pH 7 buffer. (Be sure that the buffer series is selected in the display box below the temperature box.)

Lower the electrodes into the standardization solution carefully so they do not strike the bottom of the beaker and break. If an experiment requires the use of a magnetic stirring bar in the solution whose pH is being measured, be careful that the bar does not hit the electrodes. A buffer solution is used as the standardization solution because its pH is known and it will maintain its pH in case of contamination as long as it is not excessive. A buffer solution of pH=7 is commonly used although the instrument should be standardized in the pH region of the unknown solutions.

3. Press the CAL button. When the decimal point in the pH display stops flashing, the meter is calibrated.

4. Take the electrode out of the buffer, rinse with distilled water and pat dry.

The system is now ready to measure pH values of other solutions.

Measurement of Sample pH Values

1. Place the electrode in the solution.

2. Press the READ button. When the decimal point in the pH display stops flashing, the pH of the sample is shown on the display.

3. Take the electrode out of the solution, rinse with distilled water and pat dry.
NOTE: You must rinse and dry the electrode between each sample.

4. Repeat steps 1 through 3 to measure the pH of other samples.

5. When finished, rinse the electrode with distilled water and pat dry.

6. Place the plastic cap filled with $3mol \cdot L^{-1} KCl$ back on the electrode. The electrode must be stored in the solution in this cap!

7. Turn off the meter by pressing the OFF/ON button.

Selected from "Dr. Richard A. Potts. Professor of Chemistry. The University of Michigan-Dearborn Modified by the SLC Staff"

New Words

phenomenon   *n.* 现象
voltage   *n.* 电压
electrochemical   *adj.* 电化学的
dissimilar   *adj.* 不同的，相异的
semipermeable   *adj.* 半透性的
membrane   *n.* 膜，隔膜
proportionately   *adv.* 相称地，成比例地
reference   *n.* 参考，参比
inert   *adj.* 无活动的，惰性的，迟钝的
bulb   *n.* 鳞茎，球形物
fragile   *adj.* 易碎的，脆的
electrolyte   *n.* 电解，电解液
wire   *n.* 金属丝，电线
shield   *n.* 防护物，护罩，盾，盾状物
discard   *vt.* 丢弃，抛弃；*v.* 放弃
plug   *vt.* 堵，塞，插上，插栓
self-check   自检
pat   *v.* 轻拍
rub   *v.* 擦，摩擦
contamination   *n.* 沾污，污染，污染物
magnetic   *adj.* 磁的，有磁性的，有吸引力的
decimal   *n.* 小数

Notes

1. The system consists of a pH sensor, pH Half Cell, whose voltage varies proportionately to the hydrogen ion activity of the solution, and a reference electrode, Reference Half Cell, which provides a stable and constant reference voltage.
   参考译文：该系统由一个指示电极和一个参比电极组成。指示电极又称pH半电池，其

电压随溶液里的氢离子活度不同而变化。参比电极又称参比半电池，可以提供稳定的连续参比电压。

2. The primary advantage to using a combination electrode is the ability to measure the pH of a smaller volume of sample or a sample in a container with a restricted opening.

参考译文：使用复合电极的主要优势是可以测量出微量试样或者盛装在小口容器中试样的酸碱度。

3. The main disadvantages to using a combination electrode are the more limited selection of internal elements（most have only a silver-silver chloride internal reference）and the higher cost.

参考译文：使用复合电极的主要弊端是内部元素的选择余地更小（大多数只能采用银-氯化银内参比），而且成本更高。

4. If an experiment requires the use of a magnetic stirring bar in the solution whose pH is being measured, be careful that the bar does not hit the electrodes.

参考译文：如果实验要测定带有磁性搅拌棒搅拌的溶液，千万要小心勿使搅拌棒碰到电极。

Exercises

1. Put the following into Chinese.

   electrochemical cell　　　　　　　　semipermeable membrane
   sensor　　　　　　　　　　　　　　reference electrode
   self-check　　　　　　　　　　　　magnetic stirring bar

2. Put the following into English.

   半电池　　　　　　　　　　　　　　参比电极
   蒸馏水　　　　　　　　　　　　　　玻璃电极
   酸度计　　　　　　　　　　　　　　氢离子活度

3. Questions.

   （1）What are the primary advantages of using a combination electrode?

   （2）What are the main disadvantages of using a combination electrode?

   （3）What materials must you be at a work station with if you will be using the pH meter in the laboratory?

   （4）How can you calibrate the pH meter?

   （5）What is the procedure of measuring the sample pH values?

科 技 英 语 基 本 技 巧

写作技巧：科技论文英文摘要的写作技巧

一般论文摘要多放在论文的前面。一篇好的论文摘要是全文的缩影，要求简短扼要地报道文章的主题内容，既有适当的概括性，又有一定的报道深度。因此，写好工业分析科技论文英文摘要，应注意以下几方面的问题。

1. 避免重复

不要过于简单地只把论文标题加以扩展或重复（即摘要的首句是对论文标题的扩展或重复），也不能将论文的段落标题全部加以罗列来代替摘要，使读者无法得到全文梗概。这种写法在文献中是有的，但我们要避免这种写法。

2. 篇幅要小

（1）英文摘要是具有高度概括性的一种文体，要求篇幅尽量小，句子（少用短句）尽可能少而完整。一篇摘要一般不超过 250 个单词，目的是使读者读后确定是否需要阅读全文。

（2）英文摘要应写成一段，不分段落。摘要本身要完整，能独立使用，使读者即使得不到原文，也能一目了然，获得明确的概念。因此，摘要应避免使用正文中的缩写字、图、表、化学结构式、数学公式或与脚注、插图、表格、参考文献等有关符号。允许使用公认的缩略语，但非标准的，特别是作者自构的缩略语在第一次出现时需加注全称，如：muscle protein（MP），blood plasma protein（BPP）。

3. 在文字上要遵循一般英语语法

（1）人称：摘要是从客观的角度用简练的语言介绍论文的主要内容，因此忌用第一、第二人称，而用第三人称。

（2）动态：在使用第三人称时，也应使用研究的客体作主语，这样既显得客观，也能避免叙述中心转移，从而维护统一性，又便于具体内容的陈述。采用以研究的客体作主语决定了谓语多用被动语态。

（3）谓语时态：英文摘要原则上不排斥任何时态的使用，但以一般过去时为最多见。因为摘要内叙述的工作相对于文章发表时已是过去的事，将所得的结果和结论用一般过去时表达是说明当时的情况，而不将其视为普遍性规律或真理，这样比较客观。但对规律性事物往往用一般现在时表达，因为，规律性事物过去如此，现在如此，将来也应该如此。

如在 A total of 124 batches of fresh Autrian lamb was analysed for shiga-toxin producing Escherichia coli（STEC）中，A total of 124 batches of fresh Autrian lamb 就是用研究的客体作主语，因而用的是第三人称，同时语态也是被动语态（was），而时态也很明显的是用过去时（was）。

Reading Material

2410 Series Ⅱ Nitrogen Analyzer

The PerkinElmer 2410 Series Ⅱ Nitrogen Analyzer measures nitrogen and/or protein in a wide range of materials by using an advanced combustion method. The instrument is easy-to-use with a self-contained microprocessor-controlled analyzer to deliver reliable results.

Used primarily in the food and agricultural areas, the 2410 Series Ⅱ replaces the time-consuming, multistepped Kjeldahl method of analyzing feeds, grains, cereals, dairy products, fish, meats, fruits, nuts, fertilizers, plant material, soils and sediments.

Additional application areas are in the polymer and petroleum industries. When nitrogen is the heteroelement in a polymer blend, it is monitored for quality control purposes. In the energy area, nitrogen is monitored as a contaminant or as an indication of the amount of additive present.

For virtually any material where the measurement of nitrogen and/or protein is important. the 2410 Series Ⅱ will find a place in your laboratory testing.

Standard Features

Based on the classical Dumas method, the PerkinElmer 2410 Series Ⅱ Nitrogen Analyzer employs the following design principals:

• Furnace. Operating temperatures in excess of 950℃. Combustion of sample occurs in pure oxygen environment.

• Isolation of Product Gases. Reduction of NO_x to N_2. nitrogen is separated from other combustion gases, including methane, by a gas chromatography column.

• Detection. Detection of nitrogen by a thermal conductivity detector (TCD). Detector response is automatically converted directly to weight percent nitrogen and/or protein.

• Calibration. Calibration is based on weight percent nitrogen in a single primary standard.

Principle of Operation

With the PerkinElmer 2410 Series Ⅱ Nitrogen Analyzer samples are combusted in a pure oxygen environment, with the resultant combustion gas measured in an automated fashion. The 2410 Series Ⅱ system is comprised of four major zones:

• Combustion Zone
• Gas Control Zone
• Separation Zone
• Detection Zone

In the Combustion Zone, samples encapsulated in tin are inserted automatically using an auto injector. In the presence of excess oxygen and combustion reagents, the samples are completely combusted Oxides of nitrogen are reduced to elemental nitrogen, N_2. Users have the flexibility of optimizing static and dynamic combustion conditions to meet the specific sampling need of their laboratory.

The product gas then passes through a water trap into the Gas Control Zone and is captured in the mixing chamber. Here, the gas is rapidly mixed for thorough homogenization and precisely maintained at controlled conditions of pressure, temperature and volume, which leads to highly reproducible results.

After the homogenization of the product gas, the mixing chamber is depressurized through a column in the Separation Zone of the instrument. The separation approach used is a technique known as Frontal Chromatography.

As the gas elutes, it is measured by a thermal conductivity detector in the Detection Zone of the analyzer. Since the measurement in this design is made as a steady-state, stepwise change, the variations associated with the quantification of peak signals as in other analyzers are eliminated.

Operating Gases

Unlike conventional combustion analyzers. the PerkinElmer 2410 Series Ⅱ Nitrogen Analyzer uses carbon dioxide as the carrier gas in place of the more expensive helium or oxygen gas. Moreover, the use of carbon dioxide as the carrier gas eliminates the need to remove carbon dioxide from the combustion products prior to the measurement of nitrogen. Therefore there are fewer scrubber traps to maintain.

Optimized Combustion Flexibility for Best Performance

Combustion is the most critical step to the success of the measurements and ultimately affects the accuracy and precision of the final result: the weight percent of nitrogen. The 2410 Series Ⅱ provides advanced combustion conditions of temperature, time and available oxygen. The user has the flexibility to increase the sample's combustion time in the oxygen atmosphere as well as the amount of oxygen that is introduced allowing for complete combustion of virtually any type of sample.

Gas Control Zone

A unique mixing chamber ensures a thorough mechanical homogenization of the product gas under the controlled conditions of pressure, temperature and volume. The homogeneity is important in order to achieve the most precise results. In addition, results are independent of changes in barometric pressure because the product gas is controlled to constant conditions every run.

Special Features

Optimized combustion	Offers advanced combustion conditions for static and dynamic steps. Users optimize temperature, time and available oxygen
Gas control zone	Controls pressure, temperature and volume of the product gases
Diagnostics	Monitors electronic and pneumatic components, continuously, assuring best instrument performance
Wake-up	Allows automatic instrument startup, equilibration and calibration at any operator-selected time and date
Shutdown	Allows for the automatic reduction of operating temperatures at operator-selected time and date
Gas saver	Provides automatic reduction of carrier gas flow rate with a built-in valve at an operator-selected time and date
Run counters	Aids in routine maintenance procedures and monitors reagent and expendable components

	续表
Autosampler	Pneumatic, gravity feed autoinjector with 60-position autosampler carousel. Additional samples may be added during automatic sequencing
Multiple sample mode	Up to five samples may be run; the signals of each run are used to calculate the result which is reported as a single result. In this mode, larger sample sizes may be accommodated
Advanced calculations	Selectable protein determination, calculates results on dry basis, mixtures
Automatic weight transfer	Eliminates transcription errors and simplifies operations through automatic transfer weight transfer using a PerkinElmer AD-6 ultra microbalance or TLB 60
CO_2 carrier gas	Cost effective, pre-heated carbon dioxide (99.995% pure)
Copper reagent reduction	Allows for the reduction (with 5%~8% H_2 gas mixture) of the copper reagent for reuse at operator-selected time and date
Physical details	**Power requirements** **Bench space** 100VAC±10% Width 61 cm (24 in) 120VAC±10% Depth 55 cm (22 in) 230VAC±10% Height 55 cm (22 in) Mass±45kg (100 lb)
Specifications analysis time	4 minutes
Sample size	up to 500 mg, depending on sample type and matrix. Large samples are limited by the sample matrix and content (see Analytical Range)
Resolution	1μg (10ppm)
Analytical detector range	0.001~40.0 mgs total nitrogen content
Measuring range	0.01%~100%
Accuracy	Within ±0.15% of respective theoretical values
Precision	Within a standard deviation of <0.15%

Selected from "PerkinElmer Life and Analytical Sciences 710 Bridgeport Avenue Shelton. CT 06484-4794 USA Phone: (800)762-4000 or (+1) 203-925-4602 www.perkinelmer.com"

PART 4   INDUSTRY ANALYSIS

Unit 18

Text: Sampling and Sub-sampling

Developing a Sampling Plan

In the development of a sampling procedure, the first step is to reexamine the problem definition. The analyst must review how the final results will be used and obtain an understanding of the characteristics of the sample population.

For example, the question "Has this industrial site been contaminated with Pb and Hg?" addresses the immediate concern and prompts the analyst to start thinking about other issues. This thought process would bring about additional questions, such as "Have Pb and Hg compounds been used on this site?" and "what type (s) of compounds were used?" These questions lead the analyst to an assessment of the scope of the sample population. In some cases, the whole site would be sampled. At other times, the sampling would be limited to specific areas. The analyst can then recommend a sampling procedure using methods already published on soil sampling and tailored to answering the specific question. The analyst may determine that only 3~5 samples are needed in an area where drums that contained Pb and Hg compounds were placed. The other extreme would be to proposed a random sampling over a 20 acre site, where 20 samples are pulled in a equally spaced geometric fashion. Other concerns pertain to the "normal" levels of Pb and Hg in the soil for that area where state and local environmental agencies would be contacted for baseline data.

By knowing the objective, the analyst is in a position to recommend sampling procedures that are designed to obtain a reliable answer.

A sampling plan should consider the following:
- The objective of the analyst
- The liability/cost of a wrong decision
- The accuracy and precision requirements to obtain data that will allow for a correct decision to be made
- Whether the material has been sampled before
- The degree of homogeneity
- Safety risks, site hazards, and product hazards
- The contamination risks
- The number of samples to be taken
- How the samples will be collected——by whom and in what kind of container
- Whether specific storage conditions are required

- Whether preservation is required
- Whether grinding or sieving will be used

Constructing a Sampling Program

The above information can then be used to construct a sampling program, using the following set of criteria:
- The objective as stated above
- Analyte (s) of interest and method (s) to be used
- Sampling location (s)
- Number of samples to be collected and the procedures for sampling, including sample preservation/pretreatment/storage conditions/requirements

Different Approaches to Sampling

The following approaches may be specified in the sampling plan.

Random——Random sampling means that any portion of the sample population has the same probability of being taken. Random sampling is often used for production operations that are continuous. It is also used with constraints, such as the collection of a random sample during the first, middle, and last third of a production lot that must be analyzed separately to determine if the lot is homogeneous.

Systematic——Systematic samples are collected at predetermined intervals that are defined in the sampling plan.

Stratified——Stratified sampling involves specification of depth, color, or some characteristic that must be considered in meeting the objective of the analysis.

Sequential——Sequential sampling is often used to determined if a product meets specification. Initially, samples are pulled in a "systematic" fashion and the data is evaluated. If the product is well within specification, no more sampling occurs. If the product is near the specification limit (i. e.——the mean is in specification, but the uncertainty of the measurement goes beyond the specification range), then more samples are pulled to lower the uncertainty and to determine if specification is realized.

Sub-sampling

This is the process of removing a sample aliquot for preparation and measurement from an individual sample or the aggregate sample submitted for analysis. Obtaining a representative sample is the goal and homogeneity is the primary concern, not only for the original sample collection, but the sub-sampling as well. The smaller the sample aliquot, the greater the risk of achieving sub-sampling errors that will significantly influence the accuracy of the analysis. Sub-sampling may involve an attempt to homogenize the sample by grinding, sieving, blending, or mixing the original sample.

Contamination is of particular concern when the sample is handled, ground, sieved, etc. Therefore, trace analysis should be versed on the contaminants resulting from the use of various ball mills and grinders. The use of nylon sieves will eliminate contamination risks for all metals and non-metals, except carbon.

Selected from "E877~893 (1998) Standard Practice for sample for Sampling and Sample Preparation of Iron Ores and Related Materials."

New Words

population   *n.* 群体，总体
analyst   *n.* 分析员，化验员
assessment   *n.* 评估
random   *adj.* 随机的，不规则的
geometric   *adj.* 几何的
objective   *n.* 目标，目的
homogeneity   *n.* 均匀（性）
preservation   *n.* 保存，防腐，保鲜
grind   *vt.* 研磨，磨碎
sieve   *n.* 筛
analyte   *n.* （被）分析物
stratified   *adj.* 分层的，层状的
sequential   *adj.* （按）顺序的，序列的，序贯的
specification   *n.* 技术规范，规格，（详细）说明
aliquot   *n.* 等分试样，等分部分
aggregate   *adj.* 聚集，合计
homogenize   *vt.* 使均匀，使均质
blend   *vt.* 混合，调和，搅拌
contaminate   *vt.* 污染
eliminate   *vt.* 消除，消去，脱去

Notes

1. Random sampling means that any portion of the sample population has the same probability of being taken.
   参考译文：随机采样是指样品总体被采集的概率相等。
2. Contamination is of particular concern when the sample is handled, ground, sieved, etc. Therefore, trace analysts should be versed on the contaminants resulting from the use of various ball mills and grinders. The use of nylon sieves will eliminate contamination risks for all metals and non-metals, except carbon.
   参考译文：在处理、研磨、筛分样品时应特别注意污染问题。因此，痕量化验员应熟悉因使用球磨而产生的各种污染物。使用尼龙筛子可消除所有金属和除碳以外的各种非金属受到污染的风险。

Exercises

1. Put the following into Chinese.

   sampling procedure   sample preservation

   systematic sampling   the objective of the analysis

   homogeneity   contaminant

2. Put the following into English.

   随机采样   研磨

   序贯采样   球磨机

   （被）分析物   采样地点

3. Answer the following questions.

   (1) What should a sampling plan take into consideration?

   (2) What are the criteria of constructing a sampling program?

   (3) Can you describe the approaches to sampling?

Lab Key

科技英语基本技巧

科技英语写作句型：知识

句型1. 现在已有某方面的知识

    A. there is now

         we have now (are now in possession of)

         at present we have

    B. some (detailed) knowledge of

         thorough (precise) knowledge about

    C. the gravitational field (ionizing effects, productivity) of …

         the microbial production in …

句型2. 缺乏某方面的知识

         we still know little about in transport across (the neural basis of) …

         we still do not know many features of …

         we do not know whether this is the true role of …

         we do not know brain mechanisms operate …

句型3. 从某项研究还不能获得全面的知识

    A. no complete (comprehensive) knowledge …

         no thorough knowledge of …

    B. has so far been provided by experiments wish

         was provided by experiments with …

         can be obtained from studies of …

句型 4. 已有的知识来自××研究（实验）

  A. our present knowledge of (current knowledge of …)
    some knowledge of … (of our knowledge of …)
    much of our knowledge of …
    the bulk of our knowledge
  B. comes (emerges, stems, is obtained) from
    was furnished by
    has been provided by
  C. the studies of …
    experiments with …
    the pioneer works by …

句型 5. 某方面的知识可用××

  A. knowledge of this (this phenomenon. the nature of …, the role of, the behavior of, the action of) …
    knowledge of the pathways of (the synthesis of, the details of) …
  B. is used for the determination of …
    will be used for the evaluation of …
    can hardly be used to elucidate …
    could be used to establish …

口语技巧：专业性英语口头交流常用习语

大部分专业性英语口头是通过日常对话实现的（如在办公室、教室、实验室、会议室或技术交流现场和调和安装的对话）。这种非正式对话语言中的一个最显著特征就是大量运用习惯用语。因为习语在这种对话过程中常常受到外部因素（时间、地点、人物、场合等）的影响，其含义是不可预见的，所以，必须一个一个地学。要大量掌握这种习语，确实有一定难度。但是，为了更好地以英语为工具进行专业信息交流，必须尽可能多记这类习语，将它们与书面专业术语相对照，有意识地进行替换练习，练习方式如下面例子所示：

Expression	Synonym	Example
about-face	reversal	After first rejecting our offer, she did an about-face and accepted it
back to square	At the starting point again	We're not making any progress with this approach. Let's go back to square try again

在美国科技领域中，常用的对话习语有 230 项（许多在英国英语中也常用）。

Reading Material

Sample Preparation

Selecting a Sample Preparation Method

    The selection of a preparation method is dependent upon:

- the analyte (s)
- the analyte concentration level (s)
- the sample matrix
- the instrumental measurement technique
- the required sample size

The method selected will require specific sample preparation equipment and reagents. Contamination from the atmosphere, apparatus, and reagents is a key issue when addressing the selection process.

The following checklist should be considered prior to selecting a method:
- The identity of the analytes and potential chemical forms
- The concentration range(s) of the analyte(s) and the detection limit requirement(s)
- The chemical and physical composition of the sample matrix
- The availability of apparatus and equipment
- The sample size that is available or required
- The potential for contamination during some part of the sample preparation process

Using the above information, the analyst is in a position to select the preparation technique. This involves choosing the mode of attack (acid digestion, ashing, fusion), the specific chemical reagents, and the container (s) materials needed to carry out the preparation.

Sample Preparation Techniques

1. Acid digestions of inorganic samples

A liquid that is atomized using a nebulizer is the most reliable and convenient form of sample introduction to the inductively coupled plasma. If your sample is a liquid or soluble in a liquid then the sample preparation is relatively simple. If the sample is not soluble in a solvent then sample preparation techniques such as acid digestion, ashing, or fusion are required. The solvent used exclusively for the preparation techniques to be discussed here is water.

Acid digestion have the advantage of retaining "volatile" analytes and the disadvantage of being tedious when large sample sizes are required. They are ideal if the sample size is <1 gram. The acid that is focused upon is nitric acid (HNO_3). Nitric acid is popular because of its chemical compatibility, oxidizing ability, purity, and low cost.

Nitric acid is used primarily in the preparation of inorganic sample types. It is a very useful component in the destruction of organics but cannot by itself completely decompose organic matrices.

The following is a summary of some common inorganic dissolutions using nitric acid.
- Dilute 10%～15% aqueous dilution——Alkaline earth oxides, lanthanide oxides, actinide oxides, Sc_2O_3, Y_2O_3, La_2O_3.

- $1:1\ HNO_3 + H_2O$——V_2O_5, Mn oxides, CuO, CdO, Hg oxides, Tl oxides, Pb oxidizes, Bi oxidizes.
- Concentrated (69%) HNO_3——Mn, Fe, Co, Ag, Pd, Se, As, Bi, Re.
- $1:3\ HNO_3 + HCl$ ——Pt, Au, steel, Fe/Ni alloys, Cr/Ni steel.
- $1:1:1\ HNO_3 + HF + H_2O$——The metal and oxides of Ti, Zr, Hf, Nb, W, Sn, Al, Si, Ge, Sb, Te, As, Se, Mo and numerous alloys and oxide Mixtures containing one or more of these.

2. Acid digestion of organic samples

The ability of nitric acid to react with alcohols and aromatic rings forming explosive compounds calls for caution when using nitric acid alone or in combination with other reagents in the decomposition of organic matrices. If your sample contains —OH functionality it is best to pre-treat the sample with concentrated sulfuric acid. When concentrated, the sulfuric will act as a dehydrating agent:

$$R-CH_2-CH(OH)-R' + H_2SO_4 \longrightarrow R-CH=C(OH)-R' + H_2O$$

Nitric acid is rarely used alone. It is best used in combination with sulfuric and/or perchloric acids for organic sample digestion. For samples that are not highly aromatic and/or contain a high —OH functionality, nitric acid is preferred followed by perchloric acid. The only element may be lost from a nitric/perchloric digestion is Hg.

The following are some key rules that are recommended when using nitric/perchloric acid digestions:

- Organic matrices should always be pre-treated with nitric acid.
- Perchloric acid should never be used alone.
- Perchloric acid digestions should never be allowed to go to dryness.
- Hot perchloric acid should never be added to an organic matrix.
- Sample sizes should never exceed 1 gram (dry weight for biologicals).
- Unknown organic matrices should be analyzed by molecular spectroscopy to determine primary structure before attempting the use of either nitric or perchloric acid.

Sample Preparation Procedure

Procedure Name: "Determination of Trace Metals in Biological Samples". Sample Preparation of Tissues and Serum:

(1) Place a digestion vessel holder on a 4-place analytical balance and tare. Uncap a cleaned digestion vessel and add two boiling beads.

(2) Label the digestion vessel using a graphite pencil with the assigned ID. Obtain the vital weight using a 4-place analytical balance. Record the digestion vessel weight in the analytical notebook in the space provided across from the submitter. At no time is the vessel cap to be included in the vessel weight.

(3) Use only plastic forceps, spatula, or dropping pipettes to handle the tissue and serum samples. Tissue samples are removed from their shipping container to a plastic HDPE

dish for tearing (if necessary) just prior to weighing.

(4) After recording the digestion vessel weight, tare the balance. It should read 0.0000 grams to ±0.0001 grams.

(5) Add 100μg of 1000μg/mL Yttrium internal standard to the digestion vessel. Record the weight in the analytical notebook and tare the balance.

(6) Add between 0.5 and 1.5 grams of tissue sample or 0.1 to 0.15 grams of NIST bovine liver QC sample or 2.5 to 3.0 grams of serum sample to the digestion vessel and record the sample weight to the nearest 0.0001 gram.

(7) In a acid fume hood, add 3 ml of 70% nitric acid using a disposable LDPE dropping pipette.

(8) Each group of digestions is to be accompanied with a blank and a NIST/SRM 1577b bovine liver QC sample. A group of samples consists of a full digestion block of samples. One group is equal to 24 vessels (22 samples and 2 controls). Since the NIST liver is dried, the sample weight should not exceed 0.15 grams.

(9) Place the digestion vessel in the digestion block, which should be maintained at 110℃ throughout the digestion.

(10) Brown nitrogen dioxide fumes should be observed within 5 minutes. Do not leave the digestion for the first 15 minutes. The sample should be completely dissolved within 15 minutes. Swirl the digestate to render homogeneity. With the sample weights recommended, foaming should not be a problem. If foaming does occur, remove the sample from the digestion block periodically to cool until dissolved.

(11) Continue digesting the sample with nitric acid until the brown NO_x fumes are barely visible. Place the explosion shield in front of the digestion block, put on a face shield and heavy rubber gloves. Carefully add 2 ml of 72% perchloric acid using a graduated 3 ml LDPE dropping pipette.

(12) Continue the digestion at 110℃ for 16 hours. The digestion should appear a very pale yellow to water white.

(13) Allow the digestate to cool to room temperature. Weigh the digestion vessel digestate and record this weight in the analytical notebook.

(14) Calculate the weight of the digestate and record this value in the notebook. The density of the digestate has been determined to be (1.49±0.02)g/ml. Calculate the volume of the digestate by dividing the digestate weight by the density. Enter this value in the notebook. The final volume of the digestate is brought to 10 ml using 18ml water. Calculate the ml of water to add by subtracting the digestate volume from 10 ml and enter this value in the notebook. Calculate the weight of water to add by multiplying the calculated ml by 0.997 g/ml and record this value in the notebook. Tare the analytical balance and add the calculated weight of water to the nearest ±0.02 grams. A 2-place analytical balance can be used.

(15) Mix the final sample solution after capping by hand-shaking. The sample is ready for analysis. The above procedure has been used at laboratories for processing large numbers

of biological tissues from animal feeding studies. For smaller sample numbers, a Kjeldahl digestion rack with a glass hood and caustic scrubber is more convenient and is very effective in removing any perchloric acid fumes. For samples that are harder to digest, higher temperatures reaching fumes of perchloric acid may be required.

Selected from "E877～893 (1998) Standard Practice for sample for Sampling and Sample Preparation of Iron Ores and Related Materials."

PART 5 ORGANIC ANALYSIS

Unit 19

Text: Organic Compound Identification Using Infrared Spectroscopy

A portion of the spectrum where % transmittance drops to a low value then rises back to near 100% is called a "band". A band is associated with a particular vibration within the molecule. The width of a band is described as broad or narrow based on how large a range of frequencies it covers. The efficiencies for the different vibrations determine how "intense" or strong the absorption bands are. A band is described as strong, medium, or weak depending on its depth.

In the hexane spectrum below the band for the CH stretch is strong and that for the CH bend is medium. The alkane, hexane (C_6H_{14}) gives an IR spectrum that has relatively few bands because there are only CH bonds that can stretch or bend. There are bands for CH stretches at about 3000 cm^{-1}. The CH_2 bend band appears at approximately 1450 cm^{-1} and the CH_3 bend at about 1400 cm^{-1}.

The spectrum also shows that shapes of bands can differ (see Figure 19-1).

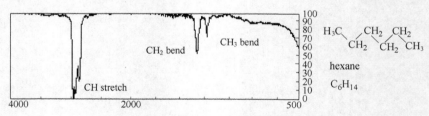

Figure 19-1 Spectrum for hexane (C_6H_{14})

Every molecule will have its own characteristic spectrum. The bands that appear depend on the types of bonds and the structure of the molecule. Study the sample spectra below, note similarities and differences, and relate these to structure and bonding within the molecules.

The spectrum for the alkene, 1-hexene, C_6H_{12}, has few strong absorption bands. The spectrum has the various CH stretch bands that all hydrocarbons show near 3000 cm^{-1}. There is a weak alkene CH stretch above 3000 cm^{-1}. H bonds on carbons 1 and 2, the two carbons that are held together by the double bond. The strong CH stretch bands below 3000 cm^{-1} come from carbon-hydrogen bonds in the CH_2 and CH_3 groups. There is an out-of-plane CH bend for the alkene in the range 1000~650 cm^{-1}. There is also an alkene CC double bond stretch at about 1650 cm^{-1} (see Figure 19-2).

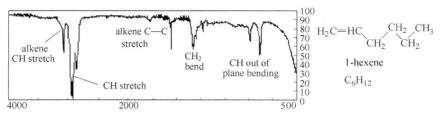

Figure 19-2 Spectrum for 1-hexene (C_6H_{12})

The spectrum for cyclohexene, (C_6H_{10}) also has few strong bands. The main band is a strong CH stretch from the CH_2 groups at about 3000 cm^{-1}. The CH stretch for the alkene CH is, as always, to the left of 3000 cm^{-1}. The CH_2 bend appears at about 1450 cm^{-1}. The other weaker bands in the range 1000～650 cm^{-1} are for the out of plane CH bending. There is a very weak alkene CC double bond stretch at about 1650 cm^{-1} (see Figure 19-3).

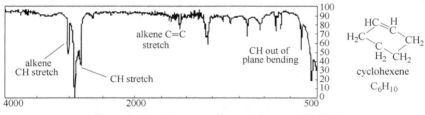

Figure 19-3 Spectrum for cyclohexene (C_6H_{10})

The IR spectrum for benzene, C_6H_6, has only four prominent bands because it is a very symmetric molecule. Every carbon has a single bond to a hydrogen. Each carbon is bonded to two other carbons and the carbon-carbon bonds are alike for all six carbons. The molecule is planar. The aromatic CH stretch appears at 3100～3000 cm^{-1}. There are aromatic CC stretch bands (for the carbon-carbon bonds in the aromatic ring) at about 1500 cm^{-1}. Two bands are caused by bending motions involving carbon-hydrogen bonds. The bands for CH bends appear at approximately 1000 cm^{-1} for the in-plane bends and at about 675 cm^{-1} for the out-of-plane bend (see Figure 19-4).

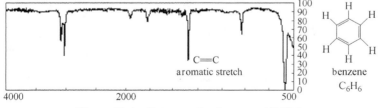

Figure 19-4 Spectrum for benzene (C_6H_6)

The IR spectrum for the alcohol, ethanol (CH_3CH_2OH), is more complicated. It has a CH stretch, an OH stretch, a CO stretch and various bending vibrations. The important point to learn here is that no matter what alcohol molecule you deal with, the OH stretch will appear as a broad band at approximately 3300～3500 cm^{-1}. Likewise the CH stretch still appears at a bout 3000 cm^{-1} (see Figure 19-5).

The spectrum for the aldehyde, octanal [$CH_3(CH_2)_6CHO$], is shown here. The most

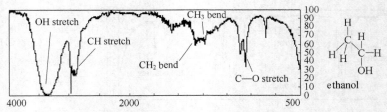

Figure 19-5　Spectrum for ethanol (CH$_3$CH$_2$OH)

important features of the spectrum are carbonyl CO stretch near 1700 cm^{-1} and the CH stretch at about 3000 cm^{-1}. If you see an IR spectrum with an intense strong band near 1700 cm^{-1} and the compound contains oxygen, the molecule most likely contains a carbonyl group, $\diagdown$C=O (see Figure 19-6).

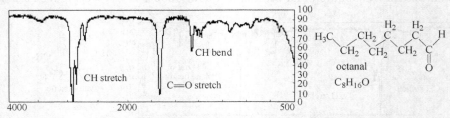

Figure 19-6　Spectrum for octanal[CH$_3$(CH$_2$)$_6$CHO]

The spectrum for the ketone, 2-pentanone, appears below. It also has a characteristic carbonyl band at 1700 cm^{-1}. The CH stretch still appears at about 3000 cm^{-1}, and the CH$_2$ bend shows up at approximately 1400 cm^{-1}. You can see the strong carbonyl CO stretch at approximately 1700 cm^{-1}. You can also see that this spectrum is different from the spectrum for octanal. At this point in your study of IR spectroscopy, you can't tell which compound is an aldehyde and which is a ketone. You can tell that both octanal and a 2-pentanone contain C—H bonds and a carbonyl group (see Figure 19-7).

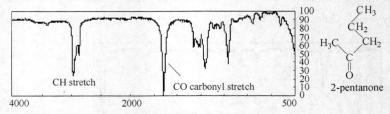

Figure 19-7　Spectrum for 2-pentanone

Carboxylic acids have spectra that are even more involved. They typically have three bands caused by bonds in the COOH functional group. The band near 1700 cm^{-1} is due to the CO double bond. The broad band centered in the range 2700~3300 cm^{-1} is caused by the presence of the OH and a band near 1400 cm^{-1} comes from the CO single bond. The spectrum for the carboxylic acid, diphenylacetic acid, appears below. Although the aromatic CH bands complicate the spectrum, you can still see the broad OH stretch between 2700~3300 cm^{-1}. It overlaps the CH stretch, which appears near 3000 cm^{-1}. A strong carbonyl CO stretch band exists near 1700 cm^{-1}. The CO single bond stretch shows up near 1200 cm^{-1} (see Figure 19-8).

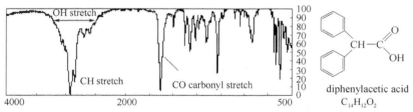

Figure 19-8   Spectrum for diphenylacetic acid

The spectrum for 1-bromobutane, C_4H_9Br, is shown here. This is relatively simple because there are only CH single bonds and the CBr bond. The CH stretch still appears at about 3000 cm^{-1}. The CH_2 bend shows up near 1400 cm^{-1}, and you can see the CBr stretch band at approximately 700 cm^{-1} (see Figure 19-9).

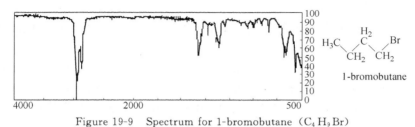

Figure 19-9   Spectrum for 1-bromobutane (C_4H_9Br)

IR spectra can be used to identify molecules by recording the spectrum for an unknown and comparing this to a library or data base of spectra of known compounds. Computerized spectra data bases and digitized spectra are used routinely in this way in research, medicine, criminology, and a number of other fields.

Selected from "Dr. Walt Volland, Bellevue Community College All rights reserved 1999, Bellevue, Washington."

New Words

infrared spectroscopy   红外光谱学，红外线分光镜
band   *n.* 波段
vibration   *n.* 振动，颤动，摇动，摆动
width   *n.* 宽度，宽广
hexane   *n.* ［化］正己烷
stretch   *v.* 伸展，伸长；*n.* 一段时间，一段路程，伸展
alkane   *n.* ［化］链烷，烷烃
cyclohexenyl   *n.* 环己烯基
plane   *n.* 平面，水平，程度；*adj.* 平的，平面的
benzene   *n.* ［化］苯
symmetric   *adj.* 对称的，均衡的

aromatic *adj.* 芬芳的，[化] 芳香族的
alcohol *n.* 酒精，酒
broad band 宽（频）带，宽波段
aldehyde *n.* [化] 醛，乙醛
carbonyl *n.* [化] 碳酰基，羰基
ketone *n.* [化] 酮
pentanone *n.* 戊酮
carboxylic *adj.* [化] 羧基的
diphenylacetic acid 二苯乙酸
overlap *v.* （与……）交叠

Notes

1. A portion of the spectrum where % transmittance drops to a low value then rises back to near 100% is called a "band".

    参考译文：透光率降到一个最低值后又重新回升到100%左右，这一段的光谱被称为"吸收谱带"。

2. If you see an IR spectrum with an intense strong band near 1700 cm^{-1} and the compound contains oxygen, the molecule most likely contains a carbonyl group.

    参考译文：如果你发现一个红外光谱在1700 cm^{-1}附近形成一个强吸收，并且该化合物含有氧元素，那么该分子最有可能含有一个羰基。

3. IR spectra can be used to identify molecules by recording the spectrum for an unknown and comparing this to a library or data base of spectra of known compounds.

    参考译文：记录一种未知化合物的光谱并将其与已知化合物的光谱进行比较，我们可以利用红外光谱鉴别分子。

Exercises

1. Put the following into Chinese.

    alkane   characteristic spectrum
    structure of the molecule   strong absorption band
    symmetric molecule   single bond
    aromatic ring   ketone
    carboxylic acid

2. Put the following into English.

    有机分析   环己烯基
    苯   宽波段
    乙醛   羰基

3. Questions.

    (1) What is a band associated with?

    (2) How many groups can a band be divided according to its depth?

    (3) Why does alkane, hexane (C_6H_{14}) gives an IR spectrum that has relatively few bands?

(4) Can you tell us the reason why the IR spectrum for benzene, C_6H_6, has only 4 prominent bands?

(5) How can we use the IR spectrum to identify molecules?

Lab Key

科 技 英 语 基 本 技 巧
科技英语写作句型：研究

句型 1. 为××摸底所进行的研究现（已）达到××程度
  A. several experimental studies
     numerous preliminary (many comprehensive) studies
     few detailed investigations
     so far no thorough (fundamental) investigations
  B. are made (now being carried out)
     were been performed
     have been undertaken (attempted, initiated)
  C. to elucidate the nature of …
     to understand the mode of action of …
     to reveal the causes of …
     in order to bring to light some facts about …

句型 2. 某项目研究的主要目的是××

(1)
  A. the chief aim
     the main purpose
     the primary object (objective)
  B. of the present study is (of our research has been)
     of this investigation was (fundamental research will be)
     of these studies will be
  C. to obtain some results which … (knowledge of …)
     to assess the role of …
     to find out whether …
     to reveal the cause of …

(2)
  A. the present (above) study
     this investigation (the study of …)
  B. was made (carried out, performed, attempted)
     has made (carried out, performed, attempted)
     has been undertaken (started, initiated, designed)

    C. with a view to elucidate (clarify, determine) …
       in order to establish (show, demonstrate) …
       to provide evidence for …
       to obtain information about …

(3)

    A. our studies were intended to
       their pioneer studies were intended to
       those earlier investigations were intended to
    B. elucidate (determine, establish) …
       provide evidence for …
       obtain information about …
       made this study …
       carried out the above study …
       performed the present study …
    C. in order to show correlation between …
       with a view to verify some basic data about …
       hoping to test the validity of the model …
       in the hope of providing new evidence of … (obtaining further information as to …)
       to the end that (so that) we (they) might bring to light some facts about …

(4)

    A. carrying out this research
       performing these studies
       undertaking the present study
    B. we hope to
       the author intended to
    C. find out … (discover, reveal, obtain) …

句型3. 某项研究的结果说明（指出，揭示）了××

(1)

    A. the studies we have performed
       these pioneer studies that authors attempted
       the studies initiated in their laboratory
       we carried out several studies which
    B. showed
       have demonstrated (indicated, suggested …)
    C. that the substance acts on
       that the effect is eliminated by …
       divers mechanisms of …

(2)

    A. the research we have done
       the investigation carried out by …

B. shows (suggests, reveals)
   has suggested (revealed)
C. a similarity between
   an increase in …
   no adaptation to … (no resistance to …)

句型 4. 从某项研究可以得出某结论
A. all these studies lead the authors to
   these and other studies to have led them to
   further research in this area leads the authors to
   (enables them to, has enabled us to)
B. conclude (suggest, believe, postulate) that …
   a conclusion that …
   a belief that

口语技巧：一般会话策略（Ⅰ）

1. 释译法

突破语音关和掌握基本口语词汇之后，在专业性英语口语练习或交谈过程中，还会经常出现不知道如何表达或找不到恰当词语的现象。在这种情况下，可以采用释义的办法进行表达。例如：指数（exponent）可以释义为 symbol that indicates what power of a factor is to be taken。这种方法可以清楚地表达自己所指的事物，克服交流障碍，避免尴尬局面。

2. 模糊法

利用一些词义相对模糊的词语表达某一概念，以弥补由于没有时间反复推敲又不可能停下来查找字典或参考资料所造成的交流缺陷。例如：In our early experiment, we inserted a temperature controlled rubidium eighty-seven filtering cell between the pumping lamp and the resonance cell. They are, eh, well, something like two separated bulbs. （在我们早期的实验中，我们把一个由温度控制的 87Rb 泡置于抽动光源与共振泡之间，这两者——嗯……就像两个隔离的灯泡。）上句中使用了句意相对模糊的 something like 这样的表达方式。比较常用的这类词语还有：a sort of…or something, It's something for…, It's a kind of thing to…。这些意义相对而言模糊的词不达意语是客观存在的。尽管它们不太精确，但在某些口头表达的特定场合，却有特殊的功能，有时还能获得较为理想的效果。

Reading Material

Organic Analysis

Organic analysis can provide a wide variety of information about organic compounds and materials such as low-molecular-weight organic compounds, polymer materials, and biochemical substances. These include qualitative and quantitative analyses, identification of functional groups, analysis of skeletal structure and sequence, analysis of trace impurities, and analysis of narrow-plane surface chemical states.

Analysis of Organic Material Composition
- Qualitative and quantitative analysis of functional groups.
- Analysis of skeletal structure of molecules.
- Qualitative and quantitative analysis of mixtures containing volatile components and low-boiling solvents.
- Analysis of nonvolatile and insoluble components.
- Analysis of inorganic elements in admixture with organic matter.

Analysis of Trace, Microscopic Substances
- Analysis of substances causing offending odors.
- Analysis of odor-causing substances in resins of containers or bottle caps.
- Analysis of trace amounts of organic substances in investigation of endocrine disrupters.
- Analysis of volatile organic compounds in the atmosphere.
- Analysis of trace impurities.
- Analysis of organic or inorganic deposits or impurities in resin, film, paint, or paper.

Analysis of Behavior When Heated
- Analysis of change in chemical structure when heated.
- Analysis of mechanism of thermal decomposition of organic material.
- Analysis of thermal reactions/thermal decomposition reactions.
- Analysis of thermal hardening behavior of thermosetting resins such as polyimide and epoxy.
- Analysis of gas generated when heated.
- Analysis of gas generated (residual solvent, organic/inorganic gases) when polymer materials are heated.
- Analysis of gas generated when adhesives or paint is heated.
- Analysis of gas component generated when various materials are heated.

Analysis of Polymer Composition
- Identification of type of polymer.
- Analysis of soluble polymer.
- Analysis of composition of refractory or insoluble polymer.
- Detailed analysis of chemical structure.
- Sequence analysis.
- Analysis of component ratio and distribution, and distribution of branching in copol-

ymers.
- Qualitative and quantitative analysis of stereoregularity, terminal groups.
- Analysis of molecular weight, molecular weight distribution.

Analysis of Additives
- Qualitative and quantitative analysis of low-molecular-weight additives in polymer.
- Analysis of high molecular weight additives in polymer.
- Analysis of inorganic filler in polymer.
- Identification of elements contained in additives.
- Determination of oligomer structure.

Analysis of Biochemical Substances
- Analysis of structure and metabolites of biopolymers, foods, and synthetic pharmaceuticals.
- Analysis of biochemical substances.
- Identification of trace proteins and analysis of post-translation modification.
- Analysis of glycoproteins and polysaccharides.
- Analysis of arrangement of amino acids in peptides and proteins.

When you analyze the spectra, it is easier if you follow a series of steps in examining each spectrum.

(1) Look first for the carbonyl $C=O$ band. Look for a strong band at $1820 \sim 1660 cm^{-1}$. This band is usually the most intense absorption band in a spectrum. It will have a medium width. If you see the carbonyl band, look for other bands associated with functional groups that contain the carbonyl by going to step 2. If no $C=O$ band is present, check for alcohols and go to step 3.

(2) If a $C=O$ is present you want to determine if it is part of an acid, an ester, or an aldehyde or ketone. At this time you may not be able to distinguish aldehyde from ketone and you will not be asked to do so.

Acid	Look for indications that an O—H is also present. It has a broad absorption near $3300 \sim 2500$ cm^{-1}. This actually will overlap the C—H stretch. There will also be a C—O single bond band near $1100 \sim 1300$ cm^{-1}. Look for the carbonyl band near $1725 \sim 1700$ cm^{-1}.
Ester	Look for C—O absorption of medium intensity near $1300 \sim 1000$ cm^{-1}. There will be no O—H band.
Aldehyde	Look for aldehyde type C—H absorption bands. These are two weak absorptions to the right of the C—H stretch near 2850 cm^{-1} and 2750 cm^{-1} and are caused by the C—H bond that is part of the CHO aldehyde functional group. Look for the carbonyl band around $1740 \sim 1720$ cm^{-1}.

Ketone	The weak aldehyde C—H absorption bands will be absent. Look for the carbonyl CO band around $1725 \sim 1705$ cm^{-1}.

(3) If no carbonyl band appears in the spectrum, look for an alcohol O—H band.

Alcohol	Look for the broad OH band near $3600 \sim 3300$ cm^{-1} and a C—O absorption band near $1300 \sim 1000$ cm^{-1}.

(4) If no carbonyl bands and no O—H bands are in the spectrum, check for double bonds, C=C, from an aromatic or an alkene.

Alkene	Look for weak absorption near 1650 cm^{-1} for a double bond. There will be a CH stretch band near 3000 cm^{-1}.
Aromatic	Look for the benzene, C=C, double bonds which appear as medium to strong absorptions in the region $1650 \sim 1450$ cm^{-1}. The CH stretch band is much weaker than in alkenes.

(5) If none of the previous groups can be identified, you may have an alkane.

Alkane	The main absorption will be the C—H stretch near 3000 cm^{-1}. The spectrum will be simple with another band near 1450 cm^{-1}.

(6) If the spectrum still cannot be assigned you may have an alkyl bromide.

Alkyl Bromide	Look for the C—H stretch and a relatively simple spectrum with an absorption to the right of 667 cm^{-1}.

Selected from "Dr. Walt Volland, Bellevue Community College All rights reserved 1999, Bellevue, Washington."

PART 6   ENVIRONMENTAL ANALYSIS

Unit 20

Text: COD is Determined by Potassium Dichromate

General Discussion

Principle: Most types of organic matter are oxidized by a boiling mixture of chromic and sulfuric acids. A sample is refluxed in strongly acid solution with a known excess of ($K_2Cr_2O_7$). After digestion, the remaining unreduced $K_2Cr_2O_7$ is titrated with ferrous ammonium sulfate to determine the amount of $K_2Cr_2O_7$ consumed and the oxidizable matter is calculated in terms of oxygen equivalent. Keep ratios of reagent weights, volumes, and strengths constant when sample volumes other than 50 ml are used. The standard 2 h reflux time may be reduced if it has been shown that a shorter period yields the same results. Some samples with very low COD or with highly heterogeneous solids content may need to be analyzed in replicate to yield the most reliable data. Results are further enhanced by reacting a maximum quantity of dichromate, provided that some residual dichromate remains.

Apparatus

- Reflux apparatus, consisting of 500 ml or 250 ml erlenmeyer flasks with ground-glass 24/40 neck and 300 mm jacket Liebig, West, or equivalent condenser with 24/40 ground-glass joint, and a hot plate having sufficient power to produce at least 1.4 W/cm^2 of heating surface, or equivalent.
- Blender.
- Pipets, Class A and wide-bore.

Reagents

- Standard potassium dichromate solution, 0.04167 mol · L^{-1}: Dissolve 12.259 g $K_2Cr_2O_7$, primary standard grade, previously dried at 150°C for 2 h, in distilled water and dilute to 1000 ml. This reagent undergoes a six-electron reduction reaction; the equivalent concentration is 6×0.04167 mol · L^{-1}.
- Sulfuric acid reagent: Add Ag_2SO_4, reagent or technical grade, crystals or powder, to H_2SO_4 at the rate of 5.5 g Ag_2SO_4/kg H_2SO_4. Let stand 1 to 2 d to dissolve. Mix.
- Ferroin indicator solution: Dissolve 1.485 g 1,10-phenanthroline monohydrate and 695 mg $FeSO_4 · 7H_2O$ in distilled water and dilute to 100 ml. This indicator solution may be

purchased already prepared.

- Standard ferrous ammonium sulfate (FAS) titrant, approximately 0.25 mol·L^{-1}: Dissolve 98 g $Fe(NH_4)_2(SO_4)_2 \cdot 6H_2O$ in distilled water. Add 20 ml H_2SO_4, cool, and dilute to 1000 ml. Standardize this solution daily against standard $K_2Cr_2O_7$ solution as follows: Dilute 25.00 ml standard $K_2Cr_2O_7$ to about 100 ml. Add 30 ml H_2SO_4 and cool. Titrate with FAS titrant using 0.10 to 0.15 ml (2 to 3 drops) ferroin indicator.

- Mercuric sulfate, $HgSO_4$, crystals or powder.

- Sulfamic acid: Required only if the interference of nitrites is to be eliminated.

- Potassium hydrogen phthalate (KHP) standard, $HOOCC_6H_4COOK$: Lightly crush and then dry KHP to constant weight at 110°C. Dissolve 425 mg in distilled water and dilute to 1000 ml. KHP has a theoretical COD of 1.176 mg $O_2 \cdot mg^{-1}$ and this solution has a theoretical COD of 500 μg $O_2 \cdot ml^{-1}$. This solution is stable when refrigerated, but not indefinitely. Be alert to development of visible biological growth. If practical, prepare and transfer solution under sterile conditions. Weekly preparation usually is satisfactory.

Procedure

Treatment of Samples with COD of $>$50 mg $O_2 \cdot L^{-1}$

Blend samples if necessary and pipet 50.00 ml into a 500 ml refluxing flask. For samples with a COD of $>$900 mg $O_2 \cdot L^{-1}$, use a smaller portion diluted to 50.00 ml. Add 1 g $HgSO_4$, several glass beads, and very slowly add 5.0 ml sulfuric acid reagent, with mixing to dissolve $HgSO_4$. Cool while mixing to avoid possible loss of volatile materials. Add 25.00 ml 0.04167 mol·L^{-1} $K_2Cr_2O_7$ solution and mix. Attach the flask to the condenser and turn on cooling water. Add remaining sulfuric acid reagent (70 ml) through open end of condenser. Continue swirling and mixing while adding sulfuric acid reagent. Caution: Mix the reflux mixture thoroughly before applying heat to prevent local heating of the flask bottom and a possible blowout of flask contents.

Cover open end of condenser with a small beaker to prevent foreign material from entering the refluxing mixture and reflux for 2 h. Cool and wash down condenser with distilled water. Disconnect reflux condenser and dilute mixture to about twice its volume with distilled water. Cool to room temperature and titrate excess $K_2Cr_2O_7$ with FAS, using 0.10 to 0.15 ml (2 to 3 drops) ferroin indicator. Although the quantity of ferroin indicator is not critical, use the same volume for all titrations. Take as the end point of the titration the first sharp color change from blue-green to reddish brown that persists for 1 min or longer. Duplicate determinations should agree within 5% of their average. Samples with suspended solids or components that are slow to oxidize may require additional determinations. The blue-green may reappear. In the same manner, reflux and titrate a blank containing the reagents and a volume of distilled water equal to that of sample.

Alternate Procedure for Low-COD Samples

Follow procedure with two exceptions: (ⅰ) use standard 0.004167 mol·L^{-1} $K_2Cr_2O_7$, and (ⅱ) titrate with standardized 0.025 mol·L^{-1} FAS. Exercise extreme care with this procedure because even a trace of organic matter on the glassware or from the atmosphere may cause gross errors. If a further increase in sensitivity is required, concentrate a larger volume of sample before digesting under reflux as follows: Add all reagents to a sample larger than 50 ml and reduce total volume to 15 ml by boiling in the refluxing flask open to the atmosphere without the condenser attached. Compute amount of $HgSO_4$ to be added (before concentration) on the basis of a weight ratio of 10∶1, $HgSO_4$∶Cl^-, using the amount of Cl^- present in the original volume of sample. Carry a blank reagent through the same procedure. This technique has the advantage of concentrating the sample without significant losses of easily digested volatile materials. Hard-to-digest volatile materials such as volatile acids are lost, but an improvement is gained over ordinary evaporative concentration methods.

Determination of Standard Solution

Evaluate the technique and quality of reagents by conducting the test on a standard potassium hydrogen phthalate solution.

Selected from "CE 432/533 Water Quality Management / Class Notes from Dr. A. T. Wallace Spring 1997."

New Words

reflux *n.* 回流，逆流，退潮
unreduced *adj.* 未减少（或缩小、减短）的
ferrous ammonium sulfate 硫酸亚铁铵
heterogeneous *adj.* 不同种类的
dichromate *n.* 重铬酸盐
digest *vi.* 消化，分解；*vt.* 消化，融会贯通
condenser *n.* 冷凝器，电容器
blender *n.* 掺和器，搅拌机
crystal *adj.* 结晶状的
1,10-phenanthroline *n.* 邻菲啰啉
monohydrate *n.* ［化］一水合物
mercuric *adj.* 水银的，含水银的，汞的
nitrite *n.* ［化］亚硝酸盐
potassium hydrogen phthalate 邻苯二甲酸氢钾
bead *n.* 珠子，水珠
volatile *adj.* 飞行的，挥发性的；*n.* 挥发物
reddish *adj.* 微红的，略带红色的
suspended solid 悬浮固体
thoroughly *adv.* 十分地，彻底地
gross *adj.* 总的，毛重的
theoretical *adj.* 理论的
evaporative *adj.* 成为蒸气的，蒸发的

Notes

1. Most types of organic matter are oxidized by a boiling mixture of chromic and sulfuric acids.

    参考译文：大多数有机物都易被沸腾的铬酸和硫酸的混合液所氧化。

2. After digestion, the remaining unreduced $K_2Cr_2O_7$ is titrated with ferrous ammonium sulfate to determine the amount of $K_2Cr_2O_7$ consumed and the oxidizable matter is calculated in terms of oxygen equivalent.

    参考译文：经过消解之后我们用硫酸亚铁铵滴定剩余的未还原的 $K_2Cr_2O_7$，来确定已经消耗掉 $K_2Cr_2O_7$ 的量以及计算可氧化物质的量，用氧当量来表示。

3. The standard 2h reflux time may be reduced if it has been shown that a shorter period yields the same results.

    参考译文：如果迹象表明时间间隔缩短后仍能产生同样效果的话，那么就可以减少 2 h 的标准回流时间。

4. Exercise extreme care with this procedure because even a trace of organic matter on the glassware or from the atmosphere may cause gross errors. If a further increase in sensitivity is required, concentrate a larger volume of sample before digesting under reflux as follows: Add all reagents to a sample larger than 50 ml and reduce total volume to 15 ml by boiling in the refluxing flask open to the atmosphere without the condenser attached.

    参考译文：运用该程序时应格外小心，这是因为即使玻璃器皿上残留或大气中一点有机物的痕迹，也会导致质量误差。如果需要进一步增加灵敏度的话，那么就在回流消解之前，我们应按如下步骤浓缩一定数量的试样：将所有的试剂滴加到大于 50ml 的试样之中，然后将试样放入敞口的回流烧瓶中煮沸将其总容积减至 15ml，不用附加冷凝器。

Exercises

1. Put the following into Chinese.

    sulfuric acid   chromic acid

    reflux apparatus   potassium dichromate

    mercuric sulfate   1,10-phenanthroline monohydrate

    gross error   blank reagent

    hard-to-digest volatile material

2. Put the following into English.

    硫酸亚铁铵   重铬酸盐

    挥发性物质   汞

    悬浮固体   邻苯二甲酸氢钾

3. Questions.

    (1) What does a reflux apparatus consist of?

    (2) How many kinds of reagents do we need to determine the COD by potassium dichromate?

（3）What is the principle behind the determination of COD by potassium dichromate?

（4）How can we go through the procedure of treating samples with COD of 50 mg $O_2 \cdot L^{-1}$?

（5）What is the advantage of alternate procedure for Low-COD sample?

Lab Key

科 技 英 语 基 本 技 巧
科技英语写作句型：方法

句型 1. 某方法已广泛应用于××
- A. the above method has
  the method we used has
  unclear magnetic resonance techniques have
- B. found a wide application（range of application）
- C. nowadays
  in biology nowadays
  in various fields
  in the examination of …

句型 2. 应用某方法进行了（完成了，获得了）××
- A. using this technique（the above procedure）
  with this method（the above technique）
- B. we（the authors, these workers）
- C. performed a few experiments which …
  carried out a number of experiments which …
  made several sets of experiments. As a result …
  have obtained certain results which …
  have succeeded in analyzing the composition of …

句型 3. 某方法在某方面（以其他方法）优越（不优越）
- A. the newly developed method
  the procedure we followed
  their technique
- B. has certain advantages over the existing methods（over the one used for, over the old procedure）
  has many advantages as compared with（over the existing methods）…
  is advantageous in some respect
  is convenient in some respect
  has no advantages over the one used for …

句型 4. 应用某方法可进行××工作

A. this method (technique)
   the above procedure …
B. allows us to demonstrate …
   enable us to observe (defect) …
   makes us to observe (defect) …
   is capable of providing (producing, revealing, detecting) …

口语技巧：一般会话策略（Ⅱ）

修正口误

在专业性英语口头表达中，免不了有时会出现讲得不太恰当或讲错的情况。这时，要及时予以解释或修正。

1. I'm afraid it is impossible to materialize the design. Well, let me put it in this way: For such a color TV system to succeed, both national and international cooperation is very necessary. 恐怕实现这种设计是不可能的——啊，让我们这样说吧，要使这种彩电系统获得成功，进行国内和国际的合作是很必要的。

2. When we speak of the developments in our world, we see what a great influence mathematics has had on them, well, instead of saying "influence", I would use the word "effect" here. 当我谈到世界上的一些发展时，我们会看到数学对它们起到了多么大的作用，这里，我使用 effect 而不用 influence。

修正口误常用的短语或句型还有：or rather, in other words, to be more exact (accurate) I mean, I was saying, excuse me 等。

Reading Material

Environmental Analysis

Environmental analysis involves the performance of chemical, physical, and biological measurements in an environmental system. This system may involve either the natural or the polluted environment, although the term "environmental analysis" is increasingly used to refer only to situations in which measurements are made of pollutants.

Normally, four types of measurements are made: qualitative analysis to identify the species present; quantitative analysis to determine how much of the species is present; speciation or characterization to establish details of chemical form and the manner in which the pollutant is actually present (such as being adsorbed onto the surface of a particle); and impact analysis in which measurements are made for the specific purpose of determining the extent to which an environmental impact is produced by the pollutants in question.

The overall objective of environmental analysis is to obtain information about both natural and pollutant species present in the environment so as to make a realistic assessment of their probable behavior. In the case of pollutants this involves assessment of their actual or potential environmental impact, which may be manifested in several ways. Thus, a pollutant species may present a toxicological hazard to plants or animals. It may also cause contamination of resources (such as air, water, and soil) so that they cannot be utilized for other purposes. The effects of pollutants on materials, especially building materials may be another area of ancient statues by sulfur dioxide in the atmosphere. A further area of environmental impact involves esthetic depreciation such as reduced visibility, dirty skies, and unpleasant odors. Finally, it is important to recognize that an environmental impact may not always be discernible by normal human perception so that detection may require sophisticated chemical or physical analysis.

In order to provide a meaningful description of the general field of environmental analysis, this field may be considered from three points of view: the basic concepts underlying the reasons for and choice of the analyses which are normally performed; the available techniques and methodology commonly used; and the current status of capabilities in environmental analysis.

Some of the philosophical concepts, which form the bases for environmental analyses, are as follows.

Purpose

Collection and analysis of an environmental sample may be undertaken for the purpose of research or monitoring, or as a spot check. A spot check analysis is used to obtain rapid information about the approximate extent or nature of an environmental problem.

Detection Limit

A statement of the detection limits, which can be attained by the analytical method being employed must always be included in providing the results of any environmental analysis. This is because considerable confusion exists about the meaning of a "zero" level of concentration. In fact, it is probably never possible to state that none of the atoms or molecules of the species in question are present so that no true zero exists.

Precision and Accuracy

In reporting an environmental analysis, it is also necessary to specify the precision and accuracy associated with the measurements. Thus, many environmental measurements involve comparison of results obtained in different systems or under different conditions (of temperature, time pollutant concentration, and so on) in the same system so that it is necessary to establish whether two numbers, which are different, are, in fact, indicative of different conditions.

State of matter in making an environmental analysis is necessary to designate the physical form of the species being analyzed. Most simply, this involves the actual state of matter in which the species exists (whether it is solid, liquid or gas) since many species (both inorganic and organic) may exist concurrently in different states. For example, certain organic gases can exist either as gases or adsorbed onto the surface of solid particles, and the analytical procedures employed for determination of each form are quite different.

Element Compound Distinction

One of the most strongly emphasized aspects of environmental analysis involves the distinction between a chemical element and the chemical compound in which that element exists. It is appropriate to establish the chemical form, in which a given element is present at a given concentration. While such a concept is philosophically acceptable, analytical methodology has not reached the stage where specification of inorganic compounds present at trace levels can readily be achieved.

Particle Surfaces

Where pollutant species are present in or associated with a condensed phase, it is sometimes necessary to establish whether the pollutant is part of the bulk system or present on its surface. Such a consideration is particularly meaningful since material present on the surface of an airborne particle, for example, comes into immediate contact with the external environment, whereas that which is distributed uniformly throughout the particle is effectively present at a much lower concentration and can exert a much lower chemical intensity at the particle surface. Since airborne particles can be inhaled, surface predominance can result in localized high concentrations of chemical species at the points of particle deposition in the lung.

Environmental Effect

The final link in the analytical/environmental effects chain involves the determination of the actual environmental effect produced. In most cases, this involves some form of biological measurement or bioassay, which determines the toxicological effect upon a biological organism. Due to the expense, complexity, and time-consuming nature of bioassays, it is usual to substitute a chemical analysis for purposes of monitoring toxicological effects, and in setting standards for compliance. In doing so, however, it is necessary to establish a dose response relationship between the level of one or more pollutant species and the degree of toxic impact.

Selected from "Warren Viessman, Jr and Mark J Hammer. Water Supply and Pollution Control, 1993."

PART 7   OIL ANALYSIS

Unit 21

Text: Oil Analysis

A detailed analysis of a sample of engine, transmission and hydraulic oils is a valuable preventive maintenance tool for an agricultural producer. In many cases it enables the identification of potential problems before a major repair is necessary, has the potential to reduce the frequencies of oil changes, and increases the resale value of used equipment.

What is Oil Analysis?

Oil analysis involves sampling and analyzing oil for various properties and materials to monitor wear and contamination in an engine, transmission or hydraulic system. Sampling and analyzing on a regular basis establishes a baseline of normal wear and can help indicate when abnormal wear or contamination is occurring.

Oil analysis works like this. Oil that has been inside any moving mechanical apparatus for a period of time reflects the exact condition of that assembly. Oil is in contact with engine or mechanical components as wear metallic trace particles enter the oil. These particles are so small they remain in suspension. Many products of the combustion process also will become trapped in the circulating oil. The oil becomes a working history of the machine.

Particles caused by normal wear and operation will mix with the oil. Any externally caused contamination also enters the oil. By identifying and measuring these impurities, you get an indication of the rate of wear and of any excessive contamination. An oil analysis also will suggest methods to reduce accelerated wear and contamination.

The typical oil analysis tests for the presence of a number of different materials to determine sources of wear, finds dirt and other contamination, and even checks for the use of appropriate lubricants.

Oil Analysis Can Detect
- Fuel dilution of lubrication oil
- Dirt contamination in the oil
- Antifreeze in the oil
- Excessive bearing wear
- Misapplication of lubricants

Some wear is normal, but abnormal levels of a particular material can give an early warning of impending problems and possibly prevent a major breakdown.

Early Detection Can
- Reduce repair bills
- Reduce catastrophic failures
- Increase machinery life
- Reduce non-scheduled downtime

Early detection with oil analysis can allow for corrective action such as repairing an air intake leak before major damage occurs. Probably one of the major advantages of an oil analysis program is being able to anticipate problems and schedule repair work to avoid downtime during a critical time of use.

Physical Tests

Some of the physical properties tested for and usually included in the analysis of an oil sample are:

Antifreeze forms a gummy substance that may reduce oil flow. It leads to high oxidation, oil thickening, high acidity, and engine failure if not corrected.

Fuel Dilution thins oil, lowers lubricating ability, and might drop oil pressure. This usually causes higher wear.

Oxidation measures gums, varnishes and oxidation products. High oxidation from oil used too hot or too long can leave sludge and varnish deposits and thicken the oil.

Total Base Mumber generally indicates the acid-neutralizing capacity still in the lubricant.

Total Solids include ash, carbon, lead salts from gasoline engines, and oil oxidation.

Viscosity is a measure of an oil's resistance to flow. Oil may thin due to shear in multi-viscosity oils or by dilution with fuel. Oil may thicken from oxidation when run too long or too hot. Oil also may thicken from contamination by antifreeze, sugar and other materials.

Metal Tests

Some of the metals tested for and usually included in the analysis of an oil sample and their potential sources are:

Aluminum (Al): Thrust washers, bearings and pistons are made of this metal. High readings can be from piston skirt scuffing, excessive ring groove wear, broken thrust washers, etc.

Boron, Magnesium, Calcium, Barium, Phosphorous, and Zinc: These metals are normally from the lubricating oil additive package. They involve detergents, dispersants, extreme-pressure additives, etc.

Chromium (Cr): Normally associated with piston rings. High levels can be caused by dirt coming through the air intake or broken rings.

Copper (Cu), Tin: These metals are normally from bearings or bushings and valve guides. Oil coolers also can contribute to copper readings along with some oil additives. In a new engine these results will normally be high during break in, but will decline in a few hundred hours.

Iron (Fe): This can come from many places in the engine such as liners, camshafts, crankshaft, valve train, timing gears, etc.

Lead (Pb): Use of regular gasoline will cause very high test results. Also associated with bearing wear, but fuel source (leaded gasoline) and sampling contamination (use of galvanized containers for sampling) are critical in interpreting this metal.

Silicon (Si): High readings generally indicate dirt or fine sand contamination from a leaking air intake system. This would act as an abrasive, causing excessive wear.

Sodium (Na): High readings of this metal normally are associated with a coolant leak, but can be from an oil additive package.

Selected from "Dominique FOMPEYDIE, Feyyaz ONUR et Pierre LEVILLAIN, Bull. Soc. Chim. France, I~2, II 5, 1982."

New Words

hydraulic   *adj.* 水力的，水压的
major repair   大修
contamination   *n.* 沾污，污染，污染物
fuel dilution   燃料稀释
antifreeze   *n.* ［化］防冻剂
misapplication   *n.* 误用，滥用
lubricant   *n.* 润滑剂
early warning   预警
anticipate   *vt.* 预期，期望；*v.* 预订，预见，可以预料
gummy   *adj.* 胶黏的，黏性的
sludge   *n.* 软泥，淤泥，矿泥，煤泥
deposit   *n.* 堆积物，沉淀物；*vt.* 存放，堆积；*vi.* 沉淀
viscosity   *n.* 黏质，黏性
resistance   *n.* 反抗，抵抗，阻抗
shear   *v.* 剪，修剪，剪切
aluminum   *n.* ［化］铝（元素符号 Al）
bearing   *n.* 轴承
piston   *n.* ［机］活塞
scuff   *vi.* 磨损；*vt.* 使磨损
boron   *n.* ［化］硼（元素符号 B）
phosphorous   *adj.* ［化］磷的（尤指含三价磷的），亚磷的
detergent   *n.* 清洁剂，去垢剂
dispersant   *n.* 分散剂
silicon   *n.* ［化］硅（元素符号 Si）
chromium   *n.* ［化］铬（元素符号 Cr）

camshaft n. 凸轮轴

lead n. [化] 铅（元素符号 Pb）

Notes

1. In many cases it enables identification of potential problems before a major repair is necessary, has the potential to reduce the frequencies of oil changes, and increases the resale value of used equipment.

    参考译文：在许多种情况下，我们进行油品分析既可提前发现隐患而不必进行大修，又有可能降低油品变化的频率，这样就可以提高旧设备转售价值。

2. Oil analysis involves sampling and analyzing oil for various properties and materials to monitor wear and contamination in an engine, transmission or hydraulic system. Sampling and analyzing on a regular basis establishes a baseline of normal wear and can help indicate when abnormal wear or contamination is occurring.

    参考译文：油品分析包括对油品采样然后分析其各种特性和所含物质，以便监测引擎系统、传输系统或液压系统的磨损和污染情况。通过对油品进行常规采样和分析，我们就可以确定一条正常磨损的基准线，一旦有异常磨损和污染情况出现时，油品分析结果就会显示出来。

3. Particles caused by normal wear and operation will mix with the oil. Any externally caused contamination also enters the oil. By identifying and measuring these impurities, you get an indication of the rate of wear and of any excessive contamination.

    参考译文：正常磨损和操作而产生的尘埃颗粒会和油混合到一起。任何表面污染物也能进入润滑油中，通过检验和测定这些杂质，可以掌握磨损比例及过度污染情况。

4. The typical oil analysis tests for the presence of a number of different materials to determine sources of wear, find dirt and other contamination, and even check for the use of appropriate lubricants.

    参考译文：典型的油品分析可以检测出油中是否出现许多不同的物质，从而可以确定磨损的原因，找到尘埃和其他的污染物，甚至还可以检测出润滑剂是否使用得当。

Exercises

1. Put the following into Chinese.

physical properties test	oxidation
high acidity	oxidation product
total solids	multi-viscosity oil
lead	boron
phosphorous	silicon

2. Put the following into English.

润滑能力	黏性
铝	镁
钡	锌

3. Questions.

(1) Why is it very important for an agricultural producer to give a detailed oil analysis?
(2) What is oil analysis? How can we proceed the process of oil analysis?
(3) By oil analysis, what can we detect?
(4) Can you point out one of the major advantages of an oil analysis program?
(5) How many physical properties are often tested for and included in an analysis of an oil sample?

Lab Key

科技英语基本技巧
科技英语写作句型：实验

句型 1. 进行××实验是为了××

(1)

A. we (the authors)

B. made (carried out) these experiments

   performed (undertook, attempted) this series of experiments

   started (initiated) this set of experiments

C. in order to show a correlation between …

   to elucidate the mechanism of …

   to evaluate the amount of …

   hoping to provide some understanding of (some information about) …

   in the hope that we might provide (elucidate, shed light on) the feature of …

(2)

A. experiments on animals (humans)

   experiments by these workers

B. where made (carried out, performed) in order to

   were started (initiated, undertook) with a view of

C. determine certain parameters

   measure the amount of …

   reveal the components of …

   obtain some data on (facts about …)

句型 2. 某项实验证实（提供）了××

A. experiments by M and his associates

   recent experiments in this area

   the present series of experiments

B. showed

   have demonstrated (indicated, suggested)

C. a much higher resistance to …
   a variety of changes in …
   that the changes are due to …

句型3. 根据实验结果可得出××结论（假设）

(1)

A. further experiments in this area
   more recent experiments in this area (of the same effect)
B. lead us (have led the authors)
   have enabled these workers
C. to conclude (believe, suggest, a conclusion)
D. that the phenomenon is related to …
   that the mechanism is put out of action

(2)

A. as a result of our experiments we concluded (came to realize)
   from out experiments the authors came to a conclusion
B. that lighting is dependent on …
   that these strains produce substances which …

口语技巧：一般会话策略（Ⅲ）

提问法：

在专业性英语会话过程中，当没有听清或听不懂对方所讲的内容时，可以酌情请对方重复或向对方提问，请求给予解释。例如，可以这样说：I beg your pardon? / Sorry, I can not follow you. / What was the last word? / Sorry, what does the word "…" mean? …这种方法尤其适用于和以英语为母语或英语水平较高的人进行交谈。这样可以有助于获得更大的语言信息，理解更多的语言材料，顺利完成整个交际活动。

总结法：

总结法即听话人用一句话或几句话概括总结一下讲话人所讲的内容，以确认自己是否完全理解对方的意图。例如，可以用下列句式来总结：You mean…? / So you are saying that…? / If I have understood correctly…运用这一技巧可使会话双方得到进一步沟通，从而更好地实现双方的交际意图。

Reading Material

Infrared Analysis as a Tool for Assessing Degradation in Used Engine Lubricants

Introduction

The use of infrared spectroscopy for routine monitoring of oil-lubricated components, breakdown products and contaminants has not been widely used in the past, although infrared studies of lubrication oils themselves have been performed for a number of years. The reason for this is that older dispersive infrared spectrometers would take several minutes to generate a spectrum of the used oil and then considerable additional time would be needed to reduce and interpret spectral data.

Oil Degradation Processes

To extract information from infrared spectra of used oils, a basic knowledge of the processes involved in oil degradation is required.

The lubricant in a combustion engine is operating in a very hostile environment, temperatures are high, and the lubricant is dispersed over a large surface area where it is exposed to chemically reactive by-products of the combustion process. In addition to this, the oil is exposed to sources of internal and external contamination.

Infrared Analysis of Used Oils

Infrared analysis on used engine oils can provide a great deal of information about what happens to an engine in service within a relatively short period of time. In a number of areas, infrared information is favoured over conventional oil analyses as it is regarded as having a greater diagnostic value. One should, however, exercise caution when interpreting data as engine design and operating conditions play an important part. A series of results from consecutive samples in which trends are evident has far greater diagnostic value than results from a lone sample. Important results that are available from infrared measurements and their interpretation are detailed below.

Soot Index—— The soot index is a measure of the level of partially burned fuel particles (soot) in the oil. The rate at which soot is deposited in the oil is dependent on engine design and operating conditions. An increase in the soot index will indicate poor combustion of the oil and filter change period that has been over-extended.

Oxidation Index——The oxidation index measures the degree to which the oil has been oxidized and is a good indicator of oil degradation. A rapid increase in oxidation may indicate engine overheating or a depletion of the anti-oxidant additive in the oil due to an over extended oil drain period.

Sulphate Index ——The sulfate index measures the extent to which sulfur-based acids have entered the oil. A rapid increase in the sulfate index could be due to depletion of oil additives, poor combustion or over-cooling.

Nitrate Index——The nitrate index measures the build-up of nitrogen compounds in the oil. These compounds cause oil thickening and deposits that interfere with lubrication. Nitration is influenced by incorrect fuel/air ratios, improper spark timing, high loads, low operating temperatures and piston-ring blow-by.

Water and Glycol——Water and Glycol may be detected at relatively low levels by FT-IR. The presence of glycol and water or glycol alone would indicate a coolant leak. Water alone does not necessarily indicate a coolant problem, as traces of water could result from condensation, if an oil sample has been taken from a cold engine.

Selected from "Nevin ERK and Feyyaz ONUR, Sci. Pharm., 64, 57~69, 1996."

PART 8 PHARMACEUTICAL ANALYSIS

Unit 22

Text: Spectrophotometric Analysis of Aspirin

Introduction

A colored complex is formed between aspirin and the iron (Ⅲ) ion. The intensity of the color is directly related to the concentration of aspirin present; therefore, spectrophotometric analysis can be used. A series of solutions with different aspirin concentrations will be prepared and complexed. The absorbance of each solution will be measured and a calibration curve will be constructed. Using the standard curve, the amount of aspirin in a commercial aspirin product can be determined.

The complex is formed by reacting the aspirin with sodium hydroxide to form the salicylate dianion.

$$\text{aspirin}(s) + OH^-(aq) \longrightarrow \text{salicylate}^{2-}(aq) + CH_3C-C^+(aq) + H_2O(l)$$

The addition of acidified iron (Ⅲ) ion produces the violet tetraaquosalicylatroiron (Ⅲ) complex.

$$\text{salicylate}^{2-} + [Fe(H_2O)_6]^{3+} + H^+ \longrightarrow [\text{Fe(H}_2\text{O})_4\text{-salicylate}]^+ + H_2O + H_3O^+$$

Purpose

The purpose of this lab is to determine the amount of aspirin in a commercial aspirin product. This lab may also be used to determine the purity of the aspirin produced in the Microscale Synthesis of Acetylsalicylic Acid lab.

Equipment/Materials

- 6~125 ml Erlenmeyer flasks
- 10 ml graduated cylinder
- 250 ml volumetric flask
- commercial aspirin product or aspirin the student has made
- acetylsalicylic acid
- 1mol · L^{-1} NaOH

- 100 ml volumetric flask
- 5 ml pipet
- 2 cuvettes
- analytical balance
- 0.02 mol·L^{-1} iron (Ⅲ) buffer
- spectrophotometer
- water

Safety
- Always wear goggles and an apron in the lab.
- Be careful while boiling the sodium hydroxide solution. NaOH solutions are dangerous, especially when hot.

Procedure

Part Ⅰ: Making Standards

(1) Mass 400 mg of acetylsalicylic acid in a 125 ml Erlenmeyer flask. Add 10 ml of a 1 mol·L^{-1} NaOH solution to the flask, and heat until the contents begin to boil.

(2) Quantitatively transfer the solution to a 250 ml volumetric flask, and dilute with distilled water to the mark.

(3) Pipet a 2.5 ml sample of this aspirin standard solution to a 50 ml volumetric flask. Dilute to the mark with a 0.02 mol·L^{-1} iron (Ⅲ) solution. Label this solution "A", and place it in a 125 ml Erlenmeyer flask.

(4) Prepare similar solutions with 2.0 ml, 1.5 ml, 1.0 ml, and 0.5 ml portions of the aspirin standard. Label these "B, C, D, and E."

Part Ⅱ: Making an unknown from a tablet

(1) Place one aspirin tablet in a 125 ml Erlenmeyer flask. Add 10 ml of a 1 mol·L^{-1} NaOH solution to the flask, and heat until the contents begin to boil.

(2) Quantitatively transfer the solution to a 250 ml volumetric flask, and dilute with distilled water to the mark.

(3) Pipet a 2.5 ml sample of this aspirin tablet solution to a 50 ml volumetric flask. Dilute to the mark with a 0.02 mol·L^{-1} iron (Ⅲ) solution. Label this solution "unknown," and place it in a 125 ml Erlenmeyer flask.

Part Ⅲ: Making an unknown from the product of the Microscale Synthesis of Acetylsalicylic Acid lab

(1) Mass all of the acetylsalicylic acid products and record the mass in the data section. Place it in a 125 ml Erlenmeyer flask. Add 10 ml of a 1 mol·L^{-1} NaOH solution to the flask, and heat until the contents begin to boil.

(2) Quantitatively transfer the solution to a 250 ml volumetric flask, and dilute with distilled water to the mark.

(3) Pipet 10 ml sample of this aspirin solution to a 50 ml volumetric flask. Dilute to the mark with a 0.02 mol·L^{-1} iron (Ⅲ) solution. [When adding 0.02 mol·L^{-1} iron (Ⅲ) solution, the color should be obviously purple. If it is not, stop adding the 0.02 mol·L^{-1} iron (Ⅲ) solution, and alert your instructor.] Label this solution "unknown", and place it

in a 125 ml Erlenmeyer flask.

Selected from "Feyyaz ONUR and Nevin ACAR, STP Pharma Sci., 5(2), 152~155, 1995."

New Words

aspirin　　*n.* 阿司匹林（解热镇痛药），乙酰水杨酸
complex　　*adj.* 复杂的，合成的，综合的；*n.* [化] 配合物
intensity　　*n.* 强烈，剧烈，强度
calibration　　*n.* 标度，刻度，校准
construct　　*vt.* 建造，构造，创立
salicylate　　*n.* [化] 水杨酸盐
violet　　*n.* 紫罗兰；*adj.* 紫罗兰色的
dianion　　*n.* 二价阴离子
acetylsalicylic acid　　[药] 乙酰水杨酸，阿司匹林
graduated cylinder　　*n.* 量筒
analytical balance　　[化] 分析天平
goggle　　*n.* 护目镜
apron　　*n.* 围裙，外表或作用类似围裙的东西，[机] 挡板，护板
lab　　*n.* 实验室，研究室
label　　*n.* 标签，签条；*vt.* 贴标签于
boil　　*n.* 沸点，沸腾；*v.* 煮沸，激动
distill　　*vt.* 蒸馏，提取；*vi.* 滴下
similar　　*adj.* 相似的，类似的
microscale　　*n.* 微小的规模，微尺度；*adj.* 微观大气现象的
synthesis　　*n.* 综合，合成

Notes

1. A colored complex is formed between aspirin and the iron (Ⅲ) ion. The intensity of the color is directly related to the concentration of aspirin present; therefore, spectrophotometric analysis can be used.

    参考译文：阿司匹林和三价铁离子形成有色配合物。其颜色的深浅与药品中所含阿司匹林的浓度有着直接关系，因此可以采用分光光度法测定阿司匹林的含量。

2. The absorbance of each solution will be measured and a calibration curve will be constructed. Using the standard curve, the amount of aspirin in a commercial aspirin product can be determined.

    参考译文：我们会测定出各个溶液的吸光度，然后将其绘制成标准曲线。对照标准曲线，我们就会确定出商用药品中阿司匹林的含量。

3. The complex is formed by reacting the aspirin with sodium hydroxide to form the salicylate dianion.

    参考译文：阿司匹林与氢氧化钠相互反应形成了水杨酸盐二价离子。这样就形成了配合物。

4. The addition of acidified iron (Ⅲ) ion produces the violet tetraaquosalicylatroiron (Ⅲ) complex.

参考译文：加入酸性的三价铁离子就会形成紫色的四水水杨酸基铁离子（Ⅲ）配合物。

5. The purpose of this lab is to determine the amount of aspirin in a commercial aspirin product. This lab may also be used to determine the purity of the aspirin produced in the Microscale Synthesis of Acetylsalicylic Acid lab.

参考译文：本实验的目的在于测定商用阿司匹林药品中的阿司匹林含量。也可以用来测定小型合成实验中生产的阿司匹林的纯度。

Exercises

1. Put the following into Chinese.

   spectrophotometric analysis   iron ion
   calibration curve   acidified iron
   volumetric flask   acetylsalicylic acid

2. Put the following into English.

   标准曲线   量筒
   分析天平   容量瓶
   分光光度计

3. Questions.

   (1) Why can we use the spectrophotometric analysis to determine the amount of aspirin?

   (2) How can the complex be formed?

   (3) What is the purpose of the Microscale Synthesis of Acetysalicylic Acid Lab?

   (4) How many materials and equipment do we need to conduct the spectrophotometric analysis of aspirin?

   (5) What is the procedure of the spectrophotometric analysis of aspirin?

Lab Key

科 技 英 语 基 本 技 巧
科技英语写作句型：理论

句型 1. 某理论是根据××提出来的

(1)

  A. our theory (their theory)
   there is a theory which
  B. is based on an assumption that the effect is …
   rests on the idea that the effect is …
   proceeds from the idea of (the principle of …, one essential principle)

(2)
    A. a new theory (another theory)
       an alterative theory
    B. is developed (proposed)
       has been worked out (proposed)
       was suggested (put forward, advanced)
    C. which is based on the assumption that …
       which proceeds from the concept of …

句型2. 根据某理论,可得出××
    A. according to B's theory
       as follows from the theory by N
       as can be seen from the theory advanced by …
    B. both processes occur simultaneously
       both effects are eliminated
       such disturbances may cause …
       such situations lead to …
       such interactions are the basis of …

句型3. 某理论已为××所进一步(充分)证明
    A. the theory by
       the investigating (very peculiar) theory
    B. is farther proved (confirmed, corroborated)
       has been sufficiently supported (substantiated, checked, tested, verified)
    C. by many workers
       by further studies of
       in experiments with …
       at present (these days)

句型4. 某理论必定(已定)被否定(拒绝)
    A. as can be seen, the above theory
       As we have seen, the above theory
    B. must be disproved (refuted)
       has to be given up (be abandoned)
       has been rejected (renounced)
    C. as incorrect (invalid, improbable)
       as totally wrong (inconsistent)

口语技巧:怎样讲得体的英语(Ⅰ)

正式与非正式场合

    在英语会话中,有时不知如何选择得体的表达方式与人交谈,如在正式场合用非正式的语体,而在非正式场合又用严肃的语体,结果听话人不是觉得莫名其妙就是感觉受到了冒犯。在英语会话中,对于一件事可根据诸多不同的因素(如时间、场合、说话对象及说话人与听话人之间的关系等）运用不同的表达方式。所有这些因素决定了我们在表达时应注意用语。学会说得体的英语至关重要。

（1）A：With great pleasure we welcome you to the Twentieth World Congress of Philosophy in the name of the American Organizing Committee.

B：Come to attend the meeting, please.

（2）A：Thank you very much for your kind invitation dated June 6, 2002, inviting me to attend the Third Bennial Meeting.

B：Thank you for inviting me to attend the meeting.

在上述两个例子中，A 句都用于正式场合，说话人与受话人之间的关系生疏。B 句则用于非正式的语言交流中，说话人与受话人的关系较亲密。如果将 A 句的表达方式用在 B 句的语言环境中，就显得很不得体。

Reading Material

Analytical Methods of Atropine

Quality Control of Antidote
Atropine
Assay (USP, 2002)

400 mg of atropine, accurately weighed, is dissolved in 50 ml of glacial acetic acid and one drop of crystal violet TS is added. This is titrated with $0.1 \text{ mol} \cdot \text{L}^{-1}$ perchloric acid to a green end point. A blank determination is performed and any necessary correction made. Each ml of $0.1 \text{mol} \cdot \text{L}^{-1}$ perchloric acid should be equivalent to 28.94 mg of $C_{17}H_{23}NO_3$

Limit of Foreign Alkaloids and Other Impurities (USP, 2002)

A solution of atropine in methanol is prepared containing 20 mg per ml, and, by quantitative dilution of a portion of this solution with methanol, a second solution of atropine is prepared containing 1 mg per ml. Then, 25 μl of the first (20 mg per ml) atropine solution, 1 μl of the second (1 mg per ml) atropine solution and 5 μl of a methanol solution of USP Atropine Sulfate RS containing 24 mg per ml is applied to a suitable thin layer chromatographic plate, coated with an 0.5 mm layer of chromatographic silica gel. The spots are allowed to dry, and the chromatogram developed in a solvent system consisting of a mixture of chloroform, acetone and diethylamine (5∶4∶1), until the solvent front has removed about three quarters of the length of the plate. The plate is removed from the developing chamber, the solvent front is marked and the solvent allowed to evaporate. The spots on the plate are located by spraying with potassium iodoplatinate TS: the R_f value of the principal spot obtained from each test solution should correspond to that obtained from the reference standard solution: no secondary spot obtained from the first atropine solution should exhibit intensity equal or greater than the principal spot obtained from the second atropine solution (0.2%).

Atropine Sulphate
Assay (USP, 2002)

1 g of atropine sulphate, accurately weighed, is dissolved in 50 ml of glacial acetic acid, then titrated with 0.1 mol·L^{-1} perchloric acid VS, determining the end point potentiometrically. A blank determination is performed and any necessary corrections made. Each ml of 0.1 mol·L^{-1} perchloric acid should be equivalent to 67.68 mg of $(C_{17}H_{23}NO_3)_2 \cdot H_2SO_4$.

Assay (BP, 2000)

0.500 g of atropine sulphate is dissolved in 30 ml of anhydrous acetic acid R, warming if necessary. The solution is cooled then titrated with 0.1 mol·L^{-1} perchloric acid and the end point determined potentiometrically. 1 ml of 0.1 mol·L^{-1} perchloric acid should be equivalent to 67.68 mg of $C_{34}H_{48}N_2O_{10}S$.

Assay (PPRC, 2000)

0.5 g of accurately weighed atropine sulfate is dissolved in a mixture of 10 ml of glacial acetic acid and 10 ml of acetic anhydride, one to two drops of crystal violet TS is added and the solution titrated with perchloric acid (0.1 mol·L^{-1}) VS until the colour is changed to pure blue. A blank determination is performed and any necessary corrections made. Each ml of perchloric acid (0.1 mol·L^{-1}) VS should be equivalent to 67.68 mg of $(C_{17}H_{23}NO_3)_2 \cdot H_2SO_4$.

Limit of other alkaloids (USP, 2000)

150 mg atropine sulphate is dissolved in 10 ml water. To 5 ml of this solution is added a few drops of platinic chloride TS: no precipitate should be formed. To the remaining 5 ml of the solution, 2 ml of 6 mol·L^{-1} ammonium hydroxide is added and shaken vigorously: a slight opalescence may develop but no turbidity should be produced.

Limit of foreign alkaloids and decomposition products (BP, 2000)

The substance should be examined by thin-layer chromatography using silica gel G R as the coating substance. Solutions should be made up as follows.

- Test solution. 0.2 g of the substance to be examined is dissolved in methanol R and diluted to 10 ml with the same solvent.
- Reference solution(a). 1ml of the test solution is diluted to 100 ml with methanol R.
- Reference solution(b). 5ml of reference solution (a) is diluted to 10 ml with met~hanol R.

To the plate is applied separately 10 μl of each solution. These are developed over a path of 10 cm using a mixture of 90 volumes of acetone R, 7 volumes of water R and 3 volumes of concentrated ammonia R. The plate is dried at 100 ℃ to 105 ℃ for 15 minutes. It is allowed to cool then sprayed with dilute potassium iodobismuthate solution R until the spots appear. Any spot in the chromatogram thus obtained with the test solution, apart from the principal spot, should not be more intense than the spot in the chromatogram obtained with reference solution (a)(1.0 %) and not more than one such spot should be more intense than the spot in the chromatogram obtained with reference solution (b)(0.5%).

Limit of apoatropine (BP, 2000)

0.10 g atropine sulphate is dissolved in 0.01 mol·L^{-1} hydrochloric acid and diluted to 100 ml with the same acid. The absorbance is determined at 245 nm. The specific absorbance should not be greater than 4.0, calculated with reference to the anhydrous substance (about 0.5%).

Atropine Sulphate Injection

Chromatographic methods of assay are described in USP, 2002 and BP, 2000. The pH of atropine sulphate injection USP, 2002 should be between 3.0 and 6.5. It should contain not more than 55.6 USP Endotoxin units per milligram of atropine sulfate and it should meet the standard requirements for USP, 2002 injections.

Selected from "Ecz. Fak. Derg. Determination of atropine sulfate in their binary mixture, 30 (4), 1~17, 2001."

PART 9 FOOD ANALYSIS

Unit 23

Text: Determination of Nitrogen in Foodstuffs by the Kjeldahl Method

Introduction

Nitrogen is one of the elements determined thousands of times a day in analytical laboratories. To determine the element as it occurs in mixtures containing ammonium salts, nitrate, or organic nitrogen compounds the method of Kjeldahl is probably most frequently used. The most critical step in this procedure is the sulfuric acid oxidation of the organic compound. During this operation the carbon in the sample is converted to carbon dioxide and the hydrogen to water. The fate of the nitrogen is highly dependent upon the form in which it occurs in the compound: amides and amines, as in proteinaceous material, is converted to ammonium ion whereas oxides of nitrogen may be loss. In the latter cases, the kjeldahl method leads to erroneously low results unless the sample is treated with a reducing agent prior to the digestion step.

Digestion of the Sample

The most time-consuming step in the kjeldahl analysis is the digestion process. One method, which is used to improve the kinetics of the process, involves the addition of a neutral salt such as potassium sulfate. This raises the boiling point of the sulfuric acid and hence the temperature at which the digestion is carried out.

Catalysts may also be used to assist the digestion process; the most common of these includes mercury, copper, selenium, or compounds of these elements. The ions of mercury or copper, if present, should be precipitated (e. g. , as their sulfides) prior to the distillation step; otherwise, some ammonia will be retained as ammine complexes of the metal ion.

Analysis of the Ammonium Salt

Neutralization provides a simple means for the determination of the ammonium salt. Three steps are involved in the process: first, the compound is decomposed with an excess of strong base; the liberated ammonia is then distilled from the mixture and collected quantitatively; and finally, the amount of ammonia is determined by neutralization titration.

The distillation apparatus is for the analysis of ammonia. The Kjeldahl flask is connected to a spray flask that serves to prevent small droplets of the strongly alkaline solution from being carried over in the vapor stream. A water-cooled condenser is also used; note that during the distillation the end of the adaptor tube extends below the surface of the acid solution in the receiving flask.

The method used here for the determination of ammonia involves placing a measured quantity of standard strong acid in the receiving flask. After the distillation is complete, the excess acid is back-titrated with a standard solution of base. An acid-base indicator with a transition in the slightly acidic range is used; an indicator with a neutral or basic transition range is not suitable because of the presence of ammonium ion

Calculation of the Percent Protein

A large number of different proteins contain very nearly the same percentage of nitrogen. The gravimetric factor for conversion of the weight of nitrogen to the weight of protein for normal mixture of serum proteins and protein in feeds is 6.25. A constant of 5.7 is preferred when wheat flour or any of its products is analyzed

Procedure

Weigh out samples of 0.25g to 2.5 g, depending upon the nitrogen content, and wrap each one in a filter paper. (The paper wrapping will prevent the sample from clinging to the neck of the flask.) Add 25 ml of concentrated sulfuric acid and one package of preweighed K_2SO_4 (10 g) and catalyst (0.3 g $CuSO_4$) to a 500 ml Kjeldahl flask. Clamp the flask in an inclined position in the hood and carefully heat the mixture until the H_2SO_4 is boiling. Continue the heating until the solution becomes colorless or assumes the color of the catalyst. If necessary, cautiously replace the acid lost by evaporation.

Remove the flame and allow the flask to cool; swirl the flask if the contents begin to solidify. Cautiously dilute with 200 ml of water and cool to room temperature under a water tap. Introduce 25 ml of 4% sodium sulfide solution in order to precipitate the copper.

Set up a distillation apparatus. Add precisely 50 ml of standard (0.1 mol·L^{-1}) HCl into the receiving flask; clamp the flask so that the tip of the adapter reaches just below the surface of the standard acid. Start the cooling water through the jacket of the condenser.

For each sample prepare a solution containing approximately 45 g of NaOH in about 75 ml of water. Cool this solution to room temperature before use. With the Kjeldahl flask tilted, slowly pour the caustic down the side of the container so that little mixing occurs with the solution in the flask. Add several pieces of granulated zinc and a small piece of litmus paper. Immediately connect the flask to the spray trap; very cautiously mix the solution by gentle swirling. After mixing is complete, the litmus paper should indicate that the solution is basic.

Immediately bring the solution to a boil and distill at a steady rate until one-half of the original solution remains. Watch the rate of heating during this period to prevent the receiver acid from being drawn back into the distillation flask. After the distillation is judged complete, lower the receiver flask until the tip of the adapter is well out of the standard acid; then remove the flame, disconnect the apparatus, and rinse the inside of condenser with a small amount of water. Disconnect the adapter and rinse it thoroughly. Add 2 drops of bromcresol green or methyl red and titrate the distillate with standard 0.1 mol·L^{-1} NaOH to the color change of the indicator.

Selected from "Synder, L.R.; Stadalius, M.A.; Quarry, M.A. Analytical Chemistry, 1983, Vol.55, pp.1412~1430."

New Words

Kjeldahl　*n.* 凯氏（测定氮）法
ammonium　*n.* ［化］铵
sulfuric acid　［化］硫酸
amide　*n.* 氨基化合物
amine　*n.* ［化］胺
proteinaceous　*adj.* 蛋白质的，似蛋白质的
digestion　*n.* 消解
neutral salt　中性盐
catalyst　*n.* 催化剂
selenium　*n.* ［化］硒（元素符号 Se）
ammine　*n.* ［化］氨络合物
Kjeldahl flask　长颈烧瓶，凯氏烧瓶
distillation　*n.* 蒸馏，蒸馏法，蒸馏物，精华，精髓
percentage　*n.* 百分数，百分率，百分比
gravimetric factor　重量分析因子
cling　*vi.* 粘紧，附着，紧贴
clamp　*n.* 夹子，夹钳；*vt.* 夹住，夹紧
solidify　*v.* （使）凝固，（使）团结，巩固
water tap　*n.* 水龙头
distillation apparatus　蒸馏器
receiving flask　接收瓶，收集瓶
cooling water　冷却水
granulated　*adj.* 颗粒状的
litmus paper　石蕊试纸

Notes

1. The fate of the nitrogen is highly dependent upon the form in which it occurs in the compound: amides and amines, as in proteinaceous material, is converted to ammonium ion whereas oxides of nitrogen may be loss.
   参考译文：氮元素是否存在，在很大程度上取决于它在化合物中的存在形式：如蛋白质中氨基化合物和胺被转化成铵离子，而氮氧化物或许会消失。
2. The ions of mercury or copper, if present, should be precipitated (e.g., as their sulfides) prior to the distillation step; otherwise, some ammonia will be retained as

ammine complexes of the metal ion.

参考译文：如果有汞离子或铜离子存在，那么我们在蒸馏之前，应该先将其沉淀下来（如以硫化物形式），否则的话，氨就会以金属离子的氨络合物的形式保留下来。

3. The method used here for the determination of ammonia involves placing a measured quantity of standard strong acid in the receiving flask. After the distillation is complete, the excess acid is back-titrated with a standard solution of base.

参考译文：我们这里所采用的测定氨水的方法是把一定数量的强酸标准溶液放入烧瓶中。在蒸馏结束之后，用碱标准溶液返滴定过量的酸。

4. Cool this solution to room temperature before use. With the Kjeldahl flask tilted, slowly pour the caustic down the side of the container so that little mixing occurs with the solution in the flask.

参考译文：使用前先将该溶液冷却至室温。将凯氏定氮烧瓶倾斜，缓缓地沿烧瓶的一侧倒出腐蚀剂，这样几乎不会和烧瓶里的溶液相混合。

Exercises

1. Put the following into Chinese.

   ammonium salt   organic nitrogen
   carbon dioxide   ammonium ion
   digestion process   boiling point
   ammine complex   concentrated sulfuric acid

2. Put the following into English.

   氮   硫酸钾
   返滴定   石蕊试纸
   接收瓶   消解

3. Questions.

   (1) What is the most frequently used way to determine the element of nitrogen in foodstuffs?

   (2) What is the most critical step in determining the element of nitrogen with the Kjeldahl method?

   (3) Do you know what is the most time-consuming step in the Kjeldahl analysis?

   (4) Why is an indicator with a neutral or basic transition range not suitable in the analysis of ammonium salt?

   (5) How can you set up a distillation apparatus?

Lab Key

科 技 英 语 基 本 技 巧
科技英语写作句型：数据

句型 1. 某数据（结果、发现）是十分重要（有说服力）的

(1)

A. the data available in literature

the results presented in the paper
the data obtained from recent studies of …
B. are (seem)
C. rather convincing
very interesting (important)
encouraging (promising, reliable)

(2)
A. what you have reported
B. is (seems)
C. quite amazing (fascinating clear, dramatic)
fairly convincing (consistent, interesting)
particularly interesting (important, striking)

句型 2. 某数据使我们得出××结论（假设）
A. the above results
B. most of these data
C. are chiefly obtained from recent studies of …
have been largely contributed by the laboratory of …
are chiefly provided by experiments with …
have been based on electron microscopic studies …

句型 3. 某数据（结果）可以从××考虑
A. we can look at (interpret)
one can consider (approach)
B. these data
C. as fully reliable (consistens with)
from a different angle (another viewpoint)
in terms of new concepts of …
in the light of new ideas about …

口语技巧：怎样讲得体的英语（Ⅱ）

五种语体的应用

对同一件事的表达远非正式和非正式这两种形式。按照美国社会语言学家 Martin Joos 的语体划分法，英语语体可分为五种。

（1）严肃语体——适用于极重要场合，说话人往往是地位较高的人。

（2）正式语体——适用于较严肃的场合，说话人与听者之间的关系是上下级关系，或者初次见面。

（3）普通语体——适用于同事之间的对话。

（4）非正式语体——适用于朋友之间交谈。

（5）亲密语体——适用于家庭成员之间或亲密朋友之间。

鉴于这五种语体的区分，在不同的交际环境必须使用不同的词语。假如会议马上就要开始，会议厅安静下来了。这时唯有坐在后排的一位年轻女士还在高声谈论，根本没有意识到大家都在看着她。应该怎样提醒她呢？

> 严肃语体:"Miss Smith must keep silent!"
> 正式语体:"Kindly stop talking now, Miss Smith."
> 普通语体:"Do you mind not talking now, Miss Smith?"
> 非正式语体:"Better not talk now, Mary."
> 亲密语体:"Darling…shh!"

Reading Material

Determine Vitamin C in Fruit Juices

Purpose

In this experiment we will determine the amount of vitamin C (ascorbic acid) in different fruit juices by titration of the juice with a solution of iodine. The iodine reacts rapidly with the vitamin C. If you have a juice you would like to analyze, bring about 125 ml to the lab. Compare at least two juices.

Principle

Ascorbic acid (vitamin C) is a mild reducing agent that reacts rapidly with triiodide. In this experiment, we will generate a known excess of I_3^- by the reaction of iodate with iodide, allow the reaction with ascorbic acid to proceed, and then back-titrate the excess I_3^- with thiosulfate.

$$C_6H_8O_6 + I_3^- \longrightarrow C_6H_6O_6 + 2H^+ + 3I^-$$
$$\text{ascorbic acid} \qquad \text{dehydroascorbic acid}$$

$$IO_3^- + 8I^- + 6H^+ \longrightarrow 3I_3^- + 3H_2O$$
$$\text{iodate} \qquad \text{triiodide}$$

$$I_3^- + 2S_2O_3^{2-} \longrightarrow 3I^- + S_4O_6^{2-}$$
$$\text{thiosulfate} \qquad \text{tetrathionate}$$

Preparation and Standardization of Thiosulfate Solution

(1) The starch indicator will be prepared for you. It was prepared by making a paste of 5 g of soluble starch and 5 mg of HgI_2 in 50 ml of water. The paste was poured into 500 ml of boiling water and boiled until clear.

(2) Prepare 0.07 mol·L^{-1} $Na_2S_2O_3$ by dissolving 3.8 g of $Na_2S_2O_3 \cdot 5H_2O$ and 0.05 g of Na_2CO_3 in 250 ml of freshly boiled water. Store this solution in a tightly capped dark bottle.

(3) Prepare 0.01 mol·L^{-1} KIO_3 by accurately weighing to 0.0001 g about 0.2500 g of solid reagent and dissolving it in distilled water in a 100 ml volumetric flask.

(4) Standardize the thiosulfate solution as follows: Pipet 1.000 ml of KIO_3 solution into a titration vessel. Add 10 ml (6 drops) of KI solution and 10 ml of 0.5 mol·L^{-1}

H_2SO_4. Immediately titrate with thiosulfate until the solution has lost almost all its color (pale yellow). Then add 1 drop of starch indicator and complete the titration. The disappearance of the deep blue color indicates the end point. Repeat the titration three more times with fresh portions of the KIO_3 solution. Calculate the average and the relative standardization for the molarity of the $Na_2S_2O_3$ solution.

Analysis of Vitamin C

(1) Weigh your vitamin tablet accurately to 0.0001 g. Dissolve the tablet in 60 ml of 0.5 mol·L^{-1} H_2SO_4, using a clean glass rod to help break up the solid. You might need to let it soak in solution for a half hour to allow it to soften. (Some of the solid binding material will not dissolve.) Quantitatively transfer the solution to a funnel with filter paper and allow the solution to pass into a 250 ml volumetric flask. After washing the filter paper with three small portions (about 5 ml) of the 0.5 mol·L^{-1} H_2SO_4 fill the volumetric flask to the mark with distilled water.

(2) Transfer 1.000 ml of the sample solution into the titration vessel. Add 0.100 ml (6 drops) of KI solution and 2.000 ml of standard KIO_3. Then titrate with standard thiosulfate as above. Add one drop of starch indicator just before the end point and finish the titration. Repeat the analysis three more times. Calculate the average and relative standard deviation for the number of milligrams of vitamin C per tablet.

Selected from "Synder, L.R.; Stadalius, M.A.; Quarry, M.A. Analytical Chemistry, 1983, Vol. 55, pp. 1412~1430."

PART 10 INSTRUMENTAL ANALYSIS

Unit 24

Text: Instrumental Analysis of Gas Hydrates Properties

Thermal analysis of gas hydrates

Thermal analysis profiles the physical property changes of a substance as a function of temperature while the substance is subjected to a controlled temperature programme. Differential scanning calorimetry (DSC) is one of the most commonly used thermal analysis methods, which has been utilized to obtain detailed equilibrium property data of gas hydrates during their solid-liquid phase transitions at a high pressure. DSC determines the difference in the amount of heat required to increase the temperature of a sample and a reference when both materials are subjected to identical temperature regimes in an environment cooled or heated at a controlled rate. It can be used to investigate the hydrate dissociation in various aqueous media including highly concentrated salt solutions, water/oil emulsions, and actual drilling fluids, at a high pressure ranging between 5 and 12 MPa. The same technique has also been used for probing the effect of inhibitors on natural gas hydrate formation and dissolution at low dosage levels. Time and/or temperature transformation profiles have been constructed for isothermal DSC data, yielding valuable information about complex hydrate nucleation and growth mechanisms. Heats of fusion and crystallisation involving ice and hydrates respectively have been identified by DSC studies. DSC is also used to quantify polymeric inhibitor and water interactions and the dependence of the interaction on the chemical structure of the polymer. The kinetics, thermodynamics, nucleation period and mode of action of a range of inhibitors on THF and trichlorofluoromethane hydrates have been determined using DSC.

MicroDSC and thermal modulated differential scanning calorimetric (TMDSC) have also been used for hydrates analyses at a greater range of working temperatures and pressures, and for the detection of processes that occur simultaneously, such as melting, lattice destruction and decomposition.

Crystallographic analysis of gas hydrates

Crystallographic methods are often used for the determination of the atomic and/or magnetic structure of materials. The analysis is dependent on the diffraction patterns emerging from a sample that is targeted by a beam of X-rays [X-ray diffraction (XRD)], neutrons (neutron diffraction), or electrons (electron diffraction). The three types of radiation/particles interact with a specimen in different ways. X-rays interact with the spatial distribution of the valence electrons which are affected by the total charge distribution of both the atomic nuclei and the surrounding electrons. Neutrons are scattered through atomic nuclei by the strong nuclear forces. It can also be contributed by the magnetic field if the magnetic moment of neutrons is non-zero. Electrons are charged particles. Their interactions with matter are influenced by both the positively charged atomic nuclei and the surrounding negatively charged electrons.

Topographic analysis of gas hydrates

Topographic analysis of gas hydrates can be carried out usingoptical microscopy (OM) and scanning electron microscopy (SEM), or atomic force microscopy (AFM). These microscopic technologies can produce a high-resolution scan of a sample surface, allowing magnified images and measuring and manipulating small samples at atomic scale. They have proved to be highly useful for studies of crystal-growth behaviour and morphology, as well as the distribution of gas hydrates phases in both synthetic and natural gas hydrates.

Size and size distribution analysis of gas hydrates

Particle size and size distribution analyser is a useful tool to monitor the nucleation/formation and growth of gas hydrates, and to determine the size and size distribution of methane hydrates during their formation in a pressurized reactor. It has also been used to monitor gas hydrate nucleation and growth processes in the presence of positively and negatively charged latex particles and kinetic inhibitors.

Selected from "Rojas, Y., & Lou, X. Asia-Pacific Journal of Chemical Engineering, Curtin University and John Wiley & Sons, Ltd. New York"

New Words

thermal analysis　热分析
gas hydrates　天然气水合物，可燃冰
differential scanning calorimetry (DSC)　差示扫描量热法
equilibrium　*n.* 平衡
aqueous　*adj.* 水合的
emulsion　*n.* 乳状液
dosage　*n.* 剂量
isothermal　*adj.* 等温的
crystallisation　*n.* 结晶
kinetics　*n.* 动力学
thermodynamics　*n.* 热力学
trichlorofluoromethane　三氯氟甲烷（CCl_3F）
thermal modulated differential scanning calorimetric (TMDSC)　热调制差示扫描量热法

lattice n. 框架的设计
crystallographic adj. 晶体的
topographic adj. 地形学上的
optical microscopy（OM） 光学显微镜
scanning electron microscopy（SEM） 扫描电子显微镜
atomic force microscopy（AFM） 原子力显微镜
latex particles 乳胶颗粒

Notes

1. Thermal analysis profiles the physical property changes of a substance as a function of temperature while the substance is subjected to a controlled temperature programme.

    参考译文：热分析将物质在程序温度控制下，将物质的物理性质变化作为温度的函数进行分析。

2. MicroDSC and thermal modulated differential scanning calorimetric（TMDSC）have also been used for hydrates analyses at a greater range of working temperatures and pressures, and for the detection of processes that occur simultaneously, such as melting, lattice destruction and decomposition.

    参考译文：MicroDSC 和热调制差示扫描量热（TMDSC）也已在更高的工作温度和压力范围内，以及检测同时发生的过程（如熔化、晶格破坏和分解）中用于水合物分析。

3. The analysis is dependent on the diffraction patterns emerging from a sample that is targeted by a beam of X-rays [X-ray diffraction（XRD）], neutrons（neutron diffraction）, or electrons（electron diffraction）.

    参考译文：该分析取决于由 X 射线束［X 射线衍射（XRD）］，中子（中子衍射）或电子（电子衍射）靶向的样品的衍射图案。

4. These microscopic technologies can produce a high-resolution scan of a sample surface, allowing magnified images and measuring and manipulating small samples at atomic scale.

    参考译文：这些显微技术可以对样品表面进行高分辨率扫描，放大图像，并在原子尺度上测量和操纵小样品。

Exercises

1. Put the following into Chinese.

    differential scanning calorimetry（DSC）　　　trichlorofluoromethane
    optical microscopy（OM）　　　thermal analysis
    scanning electron microscopy（SEM）　　　latex particles

2. Put the following into English.

    天然气水合物；可燃冰　　　热调制差示扫描量热法
    晶体的　　　热力学
    原子力显微镜　　　平衡

3. Questions.

    (1) Can you explain the thermal analysisis your own words?

(2) What is the most commonly used thermal analysis method?
(3) How many types of radiation/particles involved in crystallographic analysis? How do they interact with each other?
(4) What instruments will be used in topographic analysis?
(5) Whathas the size and size distribution analysis been used to do?

Lab Key

科 技 英 语 基 本 技 巧
科技英语翻译技巧与实践

随着当代科技发展日新月异，用英语发表的科技文献、论文、专利和标准日益增多，且这些资料具有客观性、准确性和严密性等特点，因此，科技英语这种有别于日常英语和文学英语的专用文体在现今科技发展中起着越来越重要的作用。

1. 科技英语翻译的特点

科技英语翻译的特点是专业性强、概念新、结构复杂、文字简练、陈述句型多、被动语态多、简略表达多以及复杂长句多。主要体现在以下几个方面。

① 客观性。科技文章用准确简洁的语句表达严谨而深刻的内容，翻译时不可随意改动数据、回避不易翻译的文字，不可加进译者的主观臆测。

② 规范性。科技术语，其实是早期学术界公认既定形成的，一般不要按自己的理解轻易更改。它们的含义和语言形式，一般不再变化。许多不同类型的科技文献，都有自己本身的格式，其中词语的表达也有相同的规律可循。

③ 文体特点。科技作品文体特点鲜明，因此在翻译实践中多用叙述方式说明事理，逻辑性强。

④ 语法结构规律。语法结构比较有规律，语言简洁明确，感情和修饰成分较少。大多使用完整句子，各种各样的复合句用得较多。

⑤ 现实性。科技翻译一直密切关注国家对科技领域的现实需要，十分积极地参与各项科研和生产活动。

2. 科技英语翻译的难点——逻辑思维

要想翻译好一篇科技英文原文，首先在翻译前必须弄清楚这篇文章的逻辑关系以及背景。科技文体着重于推理论证、逻辑思维。要运用判断和推理方法，理顺文章中各个词语之间、前后句子之间、上下段落之间存在的内在合乎逻辑规律的关系。但是，由于学科不断细分，各门学科领域之间互相渗透、交叉，科学与技术在发展过程中不断综合的趋势越来越突出。有时科技英语需要运用逻辑判断方法来解决某些似乎不合逻辑的语言现象，力求译文更准确。因此，依靠背景知识，透彻地掌握原文的逻辑关系是科技英语翻译中需要克服的难点。

3. 科技英语翻译的实践经验

在科研工作中，由于创新、研究等需要，几乎每天都要接触、使用、学习甚至翻译国外科技文献、先进技术的文件资料等，这就需要不只是专业翻译人员，普通的科研工作者也要掌握一些科技英语翻译的技巧，下面就来谈谈实际工作中的一些翻译技巧和经验，以供广大科研工作者在阅读学习以及翻译国外科技文献时使用。

（1）科技词汇与一般词汇的区别

科技词汇的数目是巨大的，这些词汇主要是把我们生活中常用的词汇通过合成法、转换法、派生法、缩略法、借用法以及混成法等几种方法重新组合形成的，科技词汇与一般词汇区别：①一词多义，一词多义是一般词汇的特征，而科技词汇要解决词的多义性的问题，已经不是语言水平的问题而是知识面的问题；②所指意义与内涵意义，一般而言，在翻译有关日常生活题材文章时，通常优先考虑词语的内涵意义，但是，科技文章却要求直截了当，不允许有丝毫的模棱两可，科技英语多用所指意义。

（2）科技词汇翻译方法

翻译科技词汇的表达方法主要有直译和意译两种。一般来说，科技翻译中大量使用的是直译的方法。直译方法有：移植、音译、象形译等；意译方法有推演、引申、解释等。还有综合翻译法，即部分直译，部分意译。

移植，就是这个词的各个词素按在词典中查出的意思依次翻译出来就可以了，也就是我们常说的因形见义。音译，有些词由于在汉语中没有贴切的词与它相对应，我们应想到音译法，有些用来表示新材料、新产品、新概念、新理论，或是用来表示药名、商标名称、机械设备的名称的词，也可以借助音译。象形译，是指"象译"和"形译"两种情况。"象译"是根据所指物体的形状进行翻译，看这个物体像什么，例如：H-beam 译为"工字梁"；而"形译"是根据词的形状进行翻译，这个词像个什么形状，例如：T-bend 译为"T字形接头"。推演，有些词如果照搬词典中意思已经解释不了原文的意思，需要根据原意结合原文中的具体语境和知识背景推断出这个词的意思。引申，在原文基础上用延续或扩展的方法来解释词义。科技文章有时也为了生动形象，用一些含义较深的词，也就是将这个词的具体意义引向抽象意义，如 head 的具体意义"头"引向抽象意义"做决策的地方"，因此 headquarter 就可以引申为"总部，指挥所"。解释，是一种辅助性的翻译手段，这种方法主要用于个别新出现的而意义又较抽象的术语，这时就用一句话解释原文这个词的意思，而不能直接给出对等词。

4. 科技英语翻译的发展

当今社会，无论是处于深造学习中还是在工作创新过程中，都离不开对世界范围科技发展的关注和对新科学、新技术的学习，掌握科技英语翻译的一般规则和常用技巧已经逐渐成为我们工作、学习的必需。

Reading Material

Gas Hydrates Structures and Promoters

A great number of gas molecules are known to form hydrates at high pressure and low temperature conditions. The three most commonly appearing structures in natural gas hydrates, namely cubic I, cubic II and hexagonal H structures, are displayed in Figure 24-1 The formation of a particular structure is largely dependent on the size of the trapped molecules. In this paper, we use sI, sII and sH representing the three structures respectively.

Hydrates systems based on gas molecules

Most natural gas molecules such as methane, ethane, hydrogen sulfide or carbon dioxide are small (0.4~0.55nm) and form structure sI. Larger molecules (0.6~0.7nm) such as propane, iso-butane form sII hydrates. Even larger molecules (0.8~0.9nm) such as isopentane, 2,2-dimetylbutane, methylcyclohexane and tert-butyl methyl ether form sH hydrates in the presence of small molecules such as methane. Interestingly, molecules smaller than 0.4nm including argon, krypton, xenon, oxygen, hydrogen and nitro-gen also form sII hydrates. Binary, ternary and multi-component gas systems have also been found in gas hydrates, exhibiting transitions between different structures. Even though different gases can form hydrates, the focus of this paper is on natural gas hydrates, including the hydrocarbons and some other organic molecules involved in the oil and gas industries.

Hydrate systems based on liquid molecules

Substances that are in the liquid form at room temperature and form hydrates at low temperature and atmospheric pressure, are of particular interests to many researchers as they form similar types of hydrate structures, as some gas molecules do, and can be used to study the latter without the requirement of high pressures. For example, ethylene oxide forms the sI structure type and tetrahydrofuran (THF) forms the sII structure type of hydrates at low

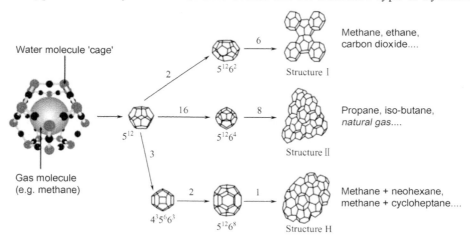

Figure 24-1 Three common hydrate unit crystal structures. 51264 indicates a water cage composed of 12 pentagonal and four hexagonal faces. The numbers over arrows indicate the number of cages of a particular type in the structure. For example, the structure I unit crystal is composed of two 512 cages, six 51262 cages. This figure is available in colour online at www.apjChemEng.com.

temperature and atmospheric pressure. THF hydrate crystals form in the presence of water or sea-water at 277.4 K and at atmospheric pressure at a molar ratio of 1 : 17 (THF to water). They form sII type structures that are usually found in natural gas hydrates and have been widely used for screening natural gas hydrate inhibitors. Other less commonly investigated liquid hydrates promoters include tetrahydropyran, chloride fluorocarbon compounds, hydrotropemolecules and some alcohols.

Selected from "Rojas, Y., & Lou, X. Asia-Pacific Journal of Chemical Engineering, Curtin University and John Wiley & Sons, Ltd. New York"

Unit 25

Text: Spectroscopic Analysis of Synthetic Lubricating Oil

The lubricating oil of engines has various important functions such as cooling engines, reducing abrasion by friction, eliminating corrosive agents. It keeps the engine in good operating condition. Lubricants consist mainly of minerals or synthetic base lubricating oil consisting of hydrocarbons paraffin, naphthenic and, to a lesser extent, aromatic hydrocarbons, to which is added a quantity of chemical additives which content is between 2% and 25%. The base oil's main function is lubrication, and additives are used to enhance this function or to provide additional properties. Control of the degradation rate of the lubricating oil and its quality enhancement can extend its life, which has both economic advantages by reducing lubricating oil consumption, and ecological advantages by reducing emissions of waste lubricant pollutants. The main causes of the degradation of the lubricating oils are related to oxidation phenomena, degradation of the molecular chains and contamination by insoluble residues and metal impurities, due to thermal phenomena and mechanical abrasions. The degradation of the lubricating oil is accompanied by changes in its viscosity. To analyze the consequences of lubricating oil aging due to thermal deterioration processes or during its use, different methods have been developed. Among these methods there are: chromatography, infrared spectroscopy, nuclear magnetic resonance, and the atomic absorption spectroscopy. For example, atomic absorption spectroscopy measurements carried out on the lubricating oil of the engine degraded during operation showed an increasing its contamination by metal from the engine or the fuel used. Determining the rate of the insoluble residues from degraded products provides information on the level of degradation of the lubricating oil.

EPR technique

Electron Paramagnetic Resonance (EPR) is a technique used to identify and determine the concentration of chemical species with unpaired electron present in solid or liquid samples. These species are free radicals, transition ions, or even defects in materials. The elementary principle of this technique is based on Zeeman effect: Under the effect of precession around itself, the electron acquires a kinetic moment and a magnetic moment of spin. The latter interacts with applied magnetic field and causes the separation of the orbital energy level into two levels of spin energy (see Figure 25-1). The following is the excitation of the electron by a radiation

$$h\nu = g\beta B$$

(ν is the excitation frequency, B is the applied magnetic field, β is the Bohr magneton, h is Planck's constant, and g is a factor that depends on the molecule). The electron absorbs this radiation and transits to the excited energy level. The EPR spectrometer analyses absorbed energy by electron. In this work, we use EPR in order to identify and follow the evolution of the free radicals produced by degradation of the lubricating oil in diesel vehicle engines.

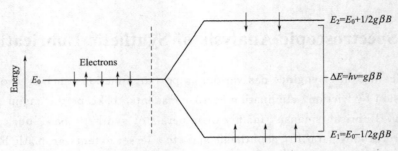

Figure 25-1 EPR transitions spin-up and spin-down electron states.

Samples preparation and EPR measurements

Samples of the degraded lubricating oil are directly taken from the engine dipstick to about every 1000km traveled by the vehicle. Thereafter, they are filtered to remove insoluble particles. After filtration, a volume of 1.256 cm^3 of lubricating oil is inserted into a quartz tube, and then analyzed by EPR spectrometer.

Electron Paramagnetic Resonance spectra reflect interactions of the sample with the external magnetic field and magnetic interactions intrinsic to the sample. To what extent these interactions contribute to the spectrum depends on the magnitude of the magnetic field and the microwave frequency. The higher the magnetic field, the larger the Zeeman contribution is. The lower the magnetic field, the more the intrinsic interactions contribute to the spectrum.

The acquisition parametersare optimized to avoid any distortion of the spectral line shape due to saturation, passage, and over-modulation. The X-band spectrum is recorded on a Magnettech MS400 spectrometer equipped with a rectangular TE102 cavity.

Acquisition parameters:
- Magnetic field swept: Centered on B0=0.3358 T
- Modulation field: 0.0003 T
- Measurement resolution: 4096
- Excitation power: 5.012 mW
- Sweeping width: 0.5 T
- Scan time: 30 min
- Number of scans: 2
- Gain: 50

The power value usedis optimized to avoid saturation, while the amplitude modulation of the magnetic field value is chosen to maintain an acceptable resolution by keeping a measurable signal for the low rates of degradation.

Fourier transforms infrared spectroscopy

The FTIR instrument used in this work is the Perkin Elmer 1725 spectrometer X. A small amount (2 μL) of the samples deposited using a Pasteur pipette between two potassium bromide discs (KBr) to obtain a thin film for the acquisition of the sample spectral data. When measured, a KBr white spectra window is collected and used as reference to calculate the sample absorbance. It is injected in virgin KBr windows to obtain its spectra in the range of 400~4000 cm^{-1}, at 4 cm^{-1} resolution.

Selected from "Zzeyani S, Mikou M, Naja J, et al., Tribology International"

New Words

corrosive   *adj.* 腐蚀性的
paraffin   *n.* 石蜡
naphthenic   *adj.* 环烷的
chromatography   *n.* 层析法
resonance   *n.* 共振
spectroscopic   *adj.* 分光镜的
synthetic   *adj.* 合成的
lubricating oil   润滑油
electron paramagnetic resonance (EPR)   电子顺磁共振
ions   *n.* [物] 离子
Zeeman effect   塞曼效应
diesel   *n.* 柴油机
dipstick   *n.* 量油计
insoluble   *adj.* 不溶的
filtration   *n.* 滤除
saturation   *n.* 饱和度
spectrum   *n.* 光谱
excitation   *n.* [物] 激发
amplitude   *n.* 振幅
Fourier transforms infrared spectroscopy   傅里叶变换红外光谱
Pasteur pipette   移液管
potassium bromide   溴化钾 (KBr)

Notes

1. Lubricants consist mainly of minerals or synthetic base lubricating oil consisting of hydrocarbons paraffin, naphthenic and, to a lesser extent, aromatic hydrocarbons, to which is added a quantity of chemical additives which content is between 2% and 25%.

    参考译文：润滑剂主要由矿物或合成基础润滑油组成，所述润滑油由烷烃以及较少程度的芳烃组成，向其中加入一定量的化学添加剂，其含量在2%~25%之间。

2. Among these methods there are: chromatography, infrared spectroscopy, nuclear magnetic resonance, and the atomic absorption spectroscopy.

    参考译文：在这些方法中有：色谱法、红外光谱法、核磁共振法和原子吸收光谱法。

3. The elementary principle of this technique is based on Zeeman effect: Under the effect of precession around itself, the electron acquires a kinetic moment and a magnetic moment

of spin.

参考译文：这种技术的基本原理是以塞曼效应为基础：在自身旋转运动的作用下，电子获得自旋的动力矩和磁矩。

4. The higher the magnetic field, the larger the Zeeman contribution is. The lower the magnetic field, the more the intrinsic interactions contribute to the spectrum.

参考译文：磁场越高，塞曼的贡献就越大。磁场越低，内在相互作用对光谱的贡献就越大。

Exercise

1. Put the following into Chinese.

   chromatography　　　　　　　lubricating oil
   saturation　　　　　　　　　spectroscopic
   excitation　　　　　　　　　Pasteur pipette

2. Put the following into English.

   石蜡　　　　　　　　　　　电子顺磁共振
   塞曼效应　　　　　　　　　过滤除
   傅里叶变换红外光谱　　　　溴化钾

3. Questions.

   （1）What are the functions of lubricating oil?
   （2）What are the components of lubricating oil?
   （3）Can you explain the Electron Paramagnetic Resonance（EPR）technique in your own words?
   （4）What are the steps of the EPR measurements? What parameters need to be acquired in the EPR measurements?
   （5）How does the Fourier transforms infrared spectroscopy acquire the sample spectral data?

Lab Key

科 技 英 语 基 本 技 巧

英文科技资料翻译的三大特点

科技英语对客观性、准确性和严密性的要求非常高，非常注重叙事逻辑上的连贯性和表达上的明晰性。科技英语力求语言的精确和表述的平和，避免旨在加强宣传效果而使用具有语言感染力的各种修辞格，防止产生行文浮华、内容虚饰等有悖于科技文体之宗旨的各类问题。英语科技文献翻译的特点综述如下。

一、专业性

译者不但要有很好的专业知识，还必须具备娴熟的双语转换技能，尤其是在专业词汇转换上。例如，element 一词，一般译为"要素""成分"，但在科技文献里却词义繁多，在化学中译作"元素",在电学中译作"电极",在无线电学中译作"元件"等。这就是科技翻译的困惑,译者一定要有很强的原文理解力,以及厚实的专业基础,才能有效地排除翻译上的疑难。不能准确地把握专业术语,译文不是让读者捧腹大笑,就是让他们不知所云。常言道,隔行如隔山。如果要翻译某个行业的科技资料,一定要事先熟悉一些基本的专业术语。例如在汽车制造业,side valve 不是指"侧面阀",而是叫"侧置气门",因为 valve 作为汽车零件不是通常意义上的"阀",而是指发动机的"气门"。若把 one of the valves in the engine must have gone wrong 译成"发动机里的一个阀门肯定出了问题",肯定会贻笑大方。译者在动手翻译之前,须掌握必要的专业知识和行业术语,比如,要知道 sump 是"油底壳",而不叫"油槽";fuel pressure pipe 是"高压油管",而不是"燃油压力管道";bearing cage 不是"轴承罩",而叫"轴承架"。简而言之,科技翻译中,专业性是第一位的。这就要求翻译人员多读科技文章,多注意科技发展趋势,多留神各个科技领域的专业术语。

二、客观性

科技文献翻译要做到不使人费解,更要做到不让人误解。英语和汉语属于两种语系,语言结构很不相同,行文叙事不相同,思维习惯更不相同。在文学作品中,翻译有较大的变通自由度,一些差异可以通过变通手段来缩小。在日常生活中,翻译亦可采用"意思对口"的释义手段来解释难译之点,如汉语日常打招呼的问话"吃过了吗?"可翻译为"Good morning!"或"Hi!"。如果真的翻译为"Have you eaten anything?"反而是可笑的。这种依照风俗习惯来转换话语意义的做法在科技表达中是不允许的。在涉及工程建筑、生产流程等有技术含量的科技翻译中,是绝对不允许的。客观是科技文献的本质,不需要过多的修饰,容不得半点儿夸张。虽然变通翻译手段的使用在科技翻译里比其在文学作品的翻译中多得多,但这种手段建立在强调客观事实的基础上,省去不译的是不需要的原文信息,翻译出来的是最新发展的科技动向。科技翻译不能像翻译文学作品那样,为了艺术上的感染力和冲击力,保全文字上的"华丽"和"流畅",而"过度"地运用"夸张"手段。这种做法对科技翻译是不可取的,有悖于其客观性表述的特征。

三、精确性

科学信息的灵魂是准确性,因此科技翻译要做到转换准确,方圆分明。这与讲究"情感"表达,追求原文意境描述的文学作品的翻译是不一样的。比如说,倍数在英汉表述和理解上容易发生偏差,造成误译。科技翻译涉及倍数关系时,理解错误必然造成翻译错误,造成企业经济上的损失,有时可能是巨大的损失。譬如,我们常常见到的"n times+形容词/副词比较级+than"这个句型,其正确的理解为"比……(大、高、重、多)$n-1$倍"。然而,有不少翻译工作者理解为"比……(大、高、重、多)n倍"。例如 Sound travels nearly three times faster in copper than in lead. 有人翻译成"声音在铜中传播的速度几乎比在铅中快三倍"。这句译文是错误的。应该译为"声音在铜中传播的速度几乎比在铅中快 2 倍"。再如,在日常生活中 every ten miles 可以笼统地理解,译成"每十英里"或"每隔九英里"。但是,在科技翻译中却有必要弄清具体的数目,才能确定 every ten miles 的精确意义。在翻译"Every ten miles a pump station will beset up."这句话时,若将 every ten miles 译成"每 10 英里",倘若这一地域总长 100 英里,就意味着要建造 10 座抽水站;如果译成"每隔九英里",就意味着要建造十一座抽水站。如何确切理解并准确翻译这个词组,需要译者在上下文中找到一个精确的答案。科学技术揭示客观事实和客观规律,是理性思维的产物。理性思维意味着严谨和精确。译者必须时时处处将理性思维体现于自己的科技翻译实践中。

Spectroscopic Analysis of Synthetic Lubricating Oil Results and Discussion

Results of EPR measurements

Analysis of EPR spectra obtained before and after the degradation of the three categories of synthetic lubricating oil

To examine the behavior of lubricating oil according to their brands and categories, we analyzed three different types of lubricants. The results of the EPR measurements on non-degraded samples are shown in (Figure 25-2).

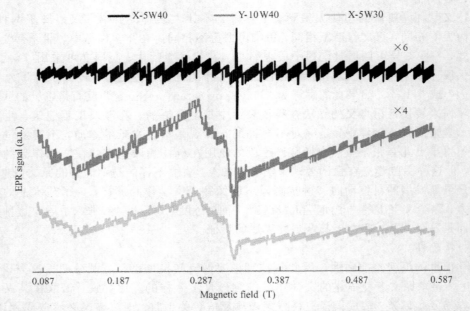

Figure 25-2 EPR spectra measured on three categories of new synthetic lubricating oil

Although the observed EPR spectra have different shapes, and their amplitudes are low, they are probably due to a low concentration of native free radicals. EPR spectra measured on samples strongly degraded are shown in (Figure 25-3). The result shows that the shape and amplitude of the spectrum depend on the nature of the lubricating oil used. Indeed, the observed spectrum of the lubricating oil X-5W40 seems wider than the other spectra measured on X-5W30 and Y-10W40 lubricating oils whose appearance is similar; it is probably due to the contribution of several free radicals. This result shows that the nature of the free radicals produced of the degradation is associated with the initial composition of the used lubricating oil.

Evolution of EPR spectra obtained during X-5W40 degradation of lubricating oil

To analyze the effect of degradation on the form and intensity of EPR spectra measured, in the interval between 0 and 10000 km traveled by the vehicle, we collected about every

PART 10　INSTRUMENTAL ANALYSIS

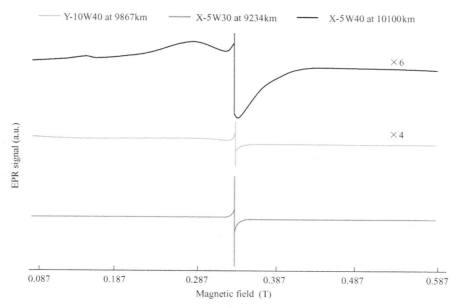

Figure 25-3　EPR spectra of degraded synthetic lubricating oil

1000 km a quantity of lubricating oil of about 20 ml from the engine. Then we performed RPE analysis under the same conditions and using the same measurement parameters. In the case of X-5W40 lubricating oil, results obtained are shown in Figure 25-3. We note that for the three oils studied, the shape of the spectra remains relatively invariable, whereas the amplitude varies regularly based on the rate of degradation. To provide a quantitative assessment of the deterioration based on the number of kilometers traveled by the vehicle, we have adopted two methods of EPR spectra analysis: method peak to peak and double integration method. The double integration method consists in integrating the experimental EPR spectrum and evaluates the area under this spectrum by performing a second integration. This last method will undoubtedly give important information on the differences in the width of the peaks observed on the three lubricating oils (Figure 25-4).

Evolution of EPR spectra as a function of lubricating oil degradation rates: Analysis by peak to peak method

Given its simplicity, the peak to peak method is most frequently used to determine the amplitude of the EPR signal. This method involves measuring the difference between the maximum value and the minimum value of the intensity of the spectrum. Thus, for the three categories of lubricating oil used, the peak to peak intensities are calculated from the measured EPR spectra.

Figure 25-5 shows the variations of these intensities depending on the number of kilometers traveled by the vehicle.

For the three categories of lubricants studied, a quasi linear evolution of the intensity of the EPR signal based on the number of kilometers traveled by the vehicle is observed. This result shows that the concentration of free radicals produced in the lubricating oil is proportional to the rate of degradation. We note that Y-10W40 lubricating oil has a much more im-

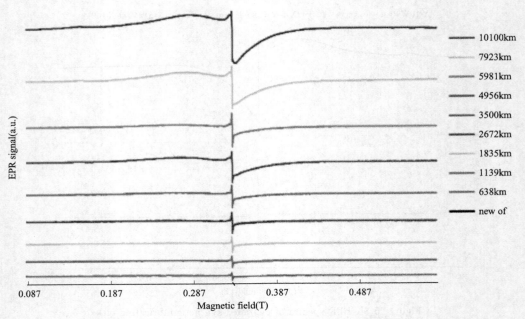

Figure 25-4 Lubricating oil X-5W40 evolution of EPR spectra

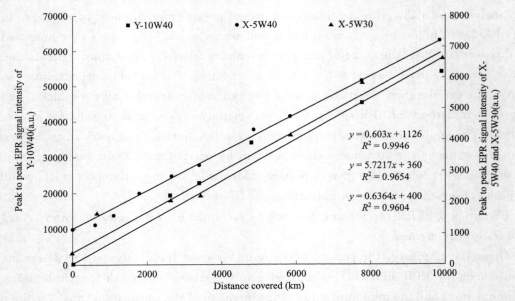

Figure 25-5 Change in peak to peak intensity of EPR signal measured on the lubricating oil samples by the number of kilometers traveled by the vehicle engines.

portant evolution of peak to peak intensity than other lubricating oils, but the line width observed in its spectrum is much less wide that seen on the X-5W40 lubricating oil. Thus, for further analysis, using the method of double integration is necessary.

Evolution of EPR spectra as a function of lubricating oil degradation rates: Analysis by the method of double integration

For a more accurate quantification of free radicals, the double integration method seems

PART 10　INSTRUMENTAL ANALYSIS

the most appropriate. This method is to integrate twice the EPR spectrum measured; the first integration brings us back to the real EPR absorption spectrum, while the second integration is to determine the area under the absorption spectrum. For an accurate assessment of the area under the spectrum, a correction of the baseline is performed (see Figure 25-6).

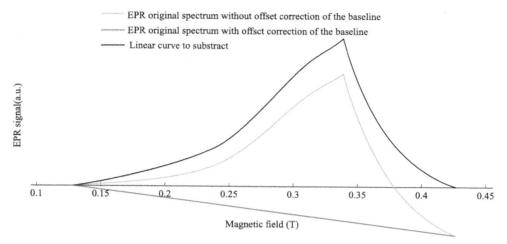

Figure 25-6　Result of the first integration of EPR spectrum measured on X-5W40 used lubricating oil (10100 km), and correction of the baseline offset after-integration.

The results of double integration are shown in (Figure 25-7). We observe that in the case of the three lubricating oils studied, the production of free radicals is almost proportional to the rate of degradation. In the case of lubricating oil X-5W40, rapid changes

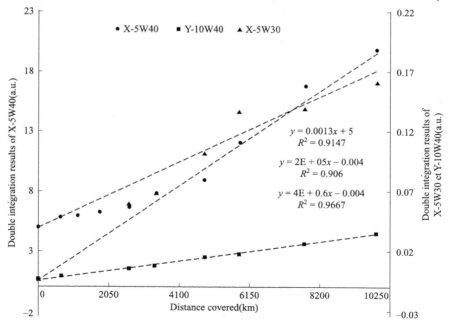

Figure 25-7　Change the results of double integration of the EPR signal intensity depending on the number of kilometers traveled by the vehicle engines.

213

in the value of double integration based on the rate of degradation are observed, while the analysis by peak to peak method reveals a relatively slow evolution. This result seems normal considering the width of the lines observed on this lubricating oil. The method of double integration also shows a substantially linear course of the degradation of lubricant depending on the distance traveled. Thus the peak to peak method is not sufficient to quantify the rate of degradation of the lubricating oil. Given these results, it is difficult to compare the results from different categories of lubricants, as each has its specific chemical composition which can influence the EPR *response*.

Aging effect analysis

To analyze aging effect, degraded lubricating oil samples are stored about one year in closed tubes at room temperature, and protected from moisture and light. Thereafter, we resumed under the same conditions as earlier, the EPR measurements on samples of degraded lubricating oil X-5W40. The results are shown in (Figure 25-8). These results indicate that the free radicals produced by the degradation of the lubricating oil are stable. Commonly, free radicals are unstable, especially when produced in a highly reactive environment. For example, moisture can promote the reactivity of free radicals and cause their instability. On the other hand, in the protected and non-reactive environments free radicals produced by thermic or mechanical phenomena or by irradiation with ionizing radiation are stable even for several months. In the case of the samples of the degraded oil, radical chain reactions are produced and lead to stable free radicals such as the peroxide radical (the de-

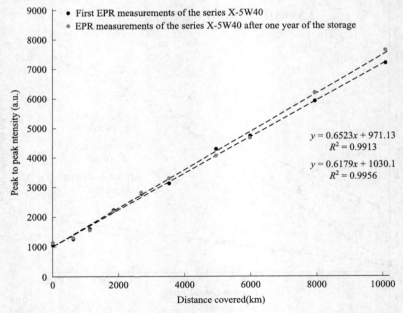

Figure 25-8　Changes in the EPR signal intensity measured on the lubricating oil samples X-5W40 depending on the number of kilometers traveled by the vehicle engines.

graded lubricant does not contain moisture since it has been subjected to high temperature). After several months of storage, the stability of the radicals resulting from the degradation is explained by the fact that the samples were stored at ambient temperature, protected from moisture and light.

Selected from "Zzeyani S, Mikou M, Naja J, et al., Tribology International"

Appendix I

Glossary

A

abrupt change 急剧（突然）变化
absorbance n. ［植、化］吸光度
absorption n. 吸收
absorption spectrum 吸收光谱，吸收频谱
accurately adv. 正确地，精确地
acetylene n. ［化］乙炔，电石气
acetylsalicylic acid ［药］乙酰水杨酸，阿司匹林
acid n. ［化］酸
adapter n. 适配器，改编者
adsorption n. 吸附
adsorption indicator 吸附指示剂
affinity n. 吸引力，亲和力
alcohol n. 酒精，酒
aldehyde n. ［化］醛，乙醛
align vi. 排列；vt. 使结盟，使成一行
alkaline adj. ［化］碱的，碱性的
alkane n ［化］链烷，烷烃
aluminum n. ［化］铝（元素符号 Al）
amide n. 氨基化合物
amine n. ［化］胺
ammine n. ［化］氨络合物
ammonia n. ［化］氨，氨水
ammonium n. ［化］铵
amount n. 数量
amplify vt. 放大，增强；v. 扩大
amplitude n. 振幅
analysis n. 分析，分解
analytical balance ［化］分析天平
angle n. ［数］角

anion n. 阴离子
anode n. ［电］阳极，正极
anticipate vt. 预期，期望；v. 预订，预见
antifreeze n. ［化］防冻剂
apparatus n. 器械，设备，仪器
approximately adv. 近似地，大约
apron n. 围裙，外表或作用类似围裙的东西，［机］挡板，护板
aqueous adj. 水的，水溶液的，水合的
aromatic adj. 芬芳的，［化］芳香族的
aspirate v. 吸气
aspirin n. 阿司匹林（解热镇痛药），乙酰水杨酸
asymmetric adj. 不均匀的，不对称的
atomic force microscopy（AFM） 原子力显微镜
atomize vt. 使分裂为原子，将……喷成雾状
audible adj. 听得见的
authentic adj. 可信的
autosampler n. 自动进样器
average n. 平均，平均水平，平均数
axis n. 轴

B

band n. 波段
bar graph 条线图，柱状图，线状光谱
base n. ［化］碱
bead n. 珠子，水珠
beaker n. 大口杯，有倾口的烧杯
beam n. 梁，桁条，（光线的）束，柱，电

波，横梁
bearing　n. 轴承
beep　n. 哔哔声；v. 嘟嘟响
Beer-Lambert law　朗伯-比耳定律
benzene　n. [化]苯
blank　n. 空白
blender　n. 掺和器，搅拌机
boil　n. 沸点，沸腾；v. 煮沸，激动
bombard　vt. 炮轰，轰击
bond　n. 化合键
boron　n. [化]硼（元素符号 B）
broad band　宽（频）带，宽波段
bromocresol green　溴甲酚绿
bromothymol blue　溴百里酚蓝
bubble　n. 泡沫
buffer solution　缓冲溶液
bulb　n. 鳞茎，球形物
bulk　n. 大小，体积，大批，大多数，散装
Bunsen burner　本生灯（即煤气灯）
buret　n. 滴定管，玻璃量管
burette　n. [化]滴定管，量管
button　n. 纽扣，[计]按钮

C

calcium　n. [化]钙（元素符号 Ca）
calculation　n. 计算，考虑
calibrate　v. 校准，使标准化，标定
calibration　n. 标度，刻度，校准
camshaft　n. 凸轮轴
capillary　n. 毛细管；adj. 毛状的，毛细作用的
carbonyl　n. [化]碳酰基，羰基
carboxylic　adj. [化]羧基的
carrier gas　n. 载气
catalyst　n. 催化剂
category　n. 种类，别，[逻]范畴

cation radical　阳离子基
charge　n. 负荷，电荷
chemical reaction　化学反应
chemical species　化学物种，化学物类
chloride　n. [化]氯化物
chromate　n. 铬酸盐
chromatography　n. 色谱，色谱法，层析法
chromium　n. [化]铬（元素符号 Cr）
clamp　n. 夹子，夹钳；vt. 夹住，夹紧
cling　vi. 粘紧，附着，紧贴
clockwise　adj. 顺时针方向的；adv. 顺时针方向地
coil　v. 盘绕，卷
colored indicator　显色指示剂
colorimeter　n. 色度计，色量计
colorless　adj. 无色的
column　n. 圆柱，柱状物，色谱柱
complex　adj. 复杂的，合成的，综合的；n. [化]配合物
component　n. 成分；adj. 组成的，构成的
concentration　n. 集中，集合，专心，浓缩，浓度
condenser　n. 冷凝器，电容器
configuration　n. 构造，结构，配置，外形
conjugate　v. 共轭，变化
construct　vt. 建造，构造，创立
contamination　n. 沾污，污染，污染物
control panel　[计]控制面板
cooling water　冷却水
copper　n. [化]铜（元素符号 Cu）
corrosive　adj. 腐蚀的，蚀坏的，有腐蚀性的；n. 腐蚀物，腐蚀剂
counterclockwise　adj. 逆时针方向的；adv. 逆时针方向地
criterion　n. 标准，规范
crystal　adj. 结晶状的

crystallisation $n.$ 结晶
crystallographic $adj.$ 晶体的
curve $n.$ 曲线，弯曲，[棒球] 曲线球，[统] 曲线图表
cuvette $n.$ 比色皿，透明小容器，试管
cyclohexenyl $n.$ 环己烯基

D

data $n.$ 数据
decimal $n.$ 小数
definable $adj.$ 可定义的
deflect $v.$ （使）偏斜，（使）偏转
deionize $v.$ [物] 除去离子，消电离
deposit $n.$ 堆积物，沉淀物；$vt.$ 存放，堆积；$vi.$ 沉淀
desired value $n.$ 期望值
detect $vt.$ 察觉，发觉，侦查，探测；$v.$ 发现
detection $n.$ 察觉，发觉，侦查，探测，发现
detector $n.$ 检测器，检波器
detergent $n.$ 清洁剂，去垢剂
dianion $n.$ 二价阴离子
dichromate $n.$ 重铬酸盐
diesel $n.$ 柴油机
differential scanning calorimetry (DSC) 差示扫描量热法
digest $vi.$ 消化，分解；$vt.$ 消化，融会贯通
digestion $n.$ 消解
digital $adj.$ 数字的，数位的，手指的；$n.$ 数字，数字式
dihydrate $n.$ [化] 二水合物
diphenylacetic acid 二苯乙酸
dipole moment 偶极矩
dipstick $n.$ 量油计
direct measurement 直接测量

discard $vt.$ 丢弃，抛弃；$v.$ 放弃
disconnect $v.$ 拆开，分离，断开
dislodge $v.$ 驱逐
disodium salt $n.$ 二钠盐
dispersant $n.$ 分散剂
dissimilar $adj.$ 不同的，相异的
dissociate $v.$ 分离，游离，分裂
distill $vt.$ 蒸馏，提取；$vi.$ 滴下
distillation $n.$ 蒸馏，蒸馏法，蒸馏物，精华，精髓
distillation apparatus 蒸馏器
distilled water 蒸馏水
distinctive $adj.$ 与众不同的，有特色的
distribute $vt.$ 分发，分配，散布；$v.$ 分发
dosage $n.$ 剂量
drain $vt.$ 排出；$vi.$ 排水，流干
dropwise $adv.$ 逐滴地，一滴一滴地
dual $adj.$ 双的，二重的，双重的

E

early warning 预警
electrical potential 电动势
electrochemical $adj.$ 电化学的
electrochemical cell 电化电池
electrochemistry $n.$ [化] 电化学
electrode $n.$ 电极
electrolyte $n.$ 电解，电解液
electron $n.$ 电子
electron bombardment 电子轰击
Electron Capture Detector (ECD) 电子捕获检测器
Electron Paramagnetic Resonance (EPR) 电子顺磁共振
element $n.$ 元素
ellipse $n.$ 椭圆，椭圆形
elute $vt.$ [化] 洗提，洗脱

employ　*vt*. 雇用，用，使用；*v*. 使用
emulsion　*n*. 乳状液
energy state　*n*. 能态
ensure　*vt*. 保证，担保，使安全，保证得到；*v*. 确保，确保，保证
equilibrium　*n*. 平衡，平静，均衡
equilibrium constant　平衡常数
equivalence　*n*. 同等，［化］等价，等值
Eriochrome Black T　铬黑T
Erlenmeyer flask　锥形（烧）瓶，爱伦美氏（烧）瓶
ethane　*n*. 乙烷
evaluation　*n*. 估价，评价，赋值
evaporate　*v*. （使）蒸发，消失
evaporative　*adj*. 成为蒸气的，蒸发的
excitation　*n*. ［物］激发
exhausted　*adj*. 耗尽的，疲惫的
experimental data　实验数据

F

faint　*adj*. 微弱的，暗淡的，模糊的
Fajans method　法扬司法
ferrous　*adj*. 铁的，含铁的，［化］亚铁的
ferrous ammonium sulfate　硫酸亚铁铵
filter　*vt*. 过滤，渗透，用过滤法除去
filtration　*n*. 过滤，筛选，滤除
flame　*n*. 火焰，光辉，光芒
Flame Ionization Detector (FID)　（氢）火焰离子化检测器
flask　*n*. 瓶，长颈瓶，细颈瓶，烧瓶，小水瓶
flexible　*adj*. 柔韧性，易曲的，灵活的，柔软的
flowmeter　*n*. 流量计
fluid　*n*. 流动性，流度；*adj*. 流动的
fluorescence　*n*. 荧光，荧光性

formula　*n*. 公式，规则，客套语，分子式
Fourier transforms infrared spectroscopy　傅立叶变换红外光谱
fragile　*adj*. 易碎的，脆的
fragment　*n*. 碎片，断片，片段
free metal　游离金属
free of　*adj*. 无……的，在……外面，摆脱……的
frequency　*n*. 频率，周率
fuel dilution　燃料稀释
functional group　官能团
fused silica　熔融石英

G

gas hydrates　天然气水合物；可燃冰
gel　*n*. 凝胶体
glassware　*n*. 玻璃器具类
goggle　*n*. 护目镜
graduated cylinder　量筒
granulated　*adj*. 颗粒状的
gravimetric factor　重量分析因子
gross　*adj*. 总的，毛重的
ground state　基态
gummy　*adj*. 胶黏的，黏性的

H

hazardous　*adj*. 危险的，冒险的，碰运气的
helium　*n*. 氦（元素符号He）
heterogeneity　*n*. 异种，异质，不同成分
heterogeneous　*adj*. 不同种类的
hexahydrate　*n*. ［化］六水合物
hexane　*n*. ［化］正己烷
hint　*n*. 暗示，提示，线索
hollow cathode lamp (HCL)　空心阴极灯
horizontal　*adj*. 地平线的，水平的

horizontal axis　水平轴（线）
hydraulic　*adj.* 水力的，水压的
hydrocarbon　*n.* 烃，碳氢化合物
hydrochloric acid　盐酸

I

identification　*n.* 辨认，鉴定，证明，视为同一
image　*n.* 图像，肖像，偶像，映像
index finger　食指
indicator　*n.* ［化］指示剂
inert　*adj.* 无活动的，惰性的，迟钝的
inert gas　*n.* 惰性气体
infrared　*adj.* 红外线的；*n.* 红外线
infrared spectroscopy　红外光谱学，红外线分光镜
ingest　*vt.* 摄取，咽下，吸收
initial　*adj.* 最初的，词首的，初始的
injection　*n.* 注射，注射剂
injector　*n.* 注射器
injury　*n.* 伤害，损坏
inlet　*n.* 进口，入口，水湾，小港，插入物
insert　*vt.* 插入，嵌入
insofar　*adv.* 在……的范围内
insoluble　*adj.* 不溶的
intensity　*n.* 强烈，剧烈，强度
interatomic distance　原子间距离
integrator　*n.* 积分仪
invert　*adj.* 转化的；*vt.* 使颠倒
ion　*n.* 离子
ion source　离子源
ionic　*adj.* 离子的
ionize　*vt.* 使离子化；*vi.* 电离
isopropanol　*n.* 异丙醇
isothermal　*adj.* 等温的

K

ketone　*n.* ［化］酮
keyboard　*n.* 键盘
kinetic energy　*n.* 动能
kinetics　*n.* 动力学
Kjeldahl　*n.* 凯氏（测定氮）法
Kjeldahl flask　长颈烧瓶，凯氏烧瓶

L

latex particles　乳胶颗粒
lattice　*n.* 框架的设计
lab　*n.* 实验室，研究室
label　*n.* 标签，签条；*vt.* 贴标签于
lead　*n.* ［化］铅（元素符号 Pb）
leak　*vi.* 漏，泄漏；*vt.* 使渗漏
legibly　*adv.* 易读地，明了地
lens　*n.* 透镜，镜头
lid　*n.* 盖子
liner　*n.* 衬垫
liquid crystal display（LCD）　*n.* 液晶显示器
litmus paper　石蕊试纸
lubricant　*n.* 润滑剂
lubricating oil　润滑油

M

magenta　*n.* 红紫色，洋红
magnetic　*adj.* 磁的，有磁性的，有吸引力的
magnetic field　磁场
maintenance　*n.* 维护，保持
major repair　大修
manganese　*n.* ［化］锰（元素符号 Mn）
manipulation　*n.* 处理，操作，操纵，被操纵
manually　*adv.* 用手，手动

manufacturer n. 制造者，厂商
mass n. 质量
mass spectrometry 质谱学，质谱分析
membrane n. 膜，隔膜
meniscus n. 新月，弯液面，凹凸透镜
mercuric adj. 水银的，含水银的，汞的
meter n. 米，计，表，仪表
methyl n. 甲基，木精
methyl group 甲基
microprocessor n. 微处理器
microscale n. 微小的规模，微尺度；adj. 微观大气现象的
milligram n. 毫克
misapplication n. 误用，滥用
mobile phase （色谱分析用的）流动相
mode n. 方式，模式，样式
modify vt. 更改，修改；v. 修改
Mohr method 莫尔法
molarity n. [化][物] 摩尔浓度
monitor vt. 监控，监测
monochromator n. 单色器
monohydrate n. [化] 一水合物
multimeter n. 万用表

N

naphthenic adj. 环烷的
nebulizer n. 喷雾器
negatively adv. 否定地，消极地
neutral adj. 中性的
neutral salt 中性盐
neutralization n. [化] 中和
nichrome wire 镍铬合金线，镍铬电热丝
nitrite n. [化] 亚硝酸盐
Nitrogen Phosphorus Detector（NPD） 氮磷检测器
nitrous oxide [化] 一氧化二氮，笑气（laughing gas）
nonlinear adj. 非线性的
noticeable adj. 显而易见的，值得注意的
nuclear magnetic resonance（NMR） 核磁共振
nut n. 坚果，螺母，螺帽

O

optical adj. 光学的
optical microscopy（OM） 光学显微镜
optimize vt. 使最优化
organic adj. 有机的
oscillate v. 振荡
oscillating electric field 振荡电场
overlap v. （与……）交叠
oxidation n. [化] 氧化
oxidize v. （使）氧化

P

paraffin n. 石蜡
parallax error 视差，判读误差
parameter n. 参数，参量
Pasteur pipette 移液管
pat v. 轻拍
path length 路径长度
pentanone n. 戊酮
percentage n. 百分数，百分率，百分比
periodate n. [化] 高碘酸盐
PerkinElmer （美国）珀金埃尔默仪器有限公司
permanganate n. [化] 高锰酸盐
persistent adj. 持久稳固的
1,10-phenanthroline n. 邻菲啰啉
phenol red 酚红
phenolphthalein n. [化] 酚酞
phenomenon n. 现象

phosphoric *adj.* ［化］磷的（尤指含五价磷的），含磷的

phosphorous *adj.* ［化］磷的（尤指含三价磷的），亚磷的

photocell *n.* 光电池

photon *n.* ［物］光子

pink *n.* 粉红色；*adj.* 粉红的

pipet *n.* 吸量管，移液管

piston *n.* ［机］活塞

plane *n.* 平面，水平，程度；*adj.* 平的，平面的

platinum wire 铂丝，白金丝

plotting *n.* 测绘，标图

plug *vt.* 堵，塞，插上，插栓

plus *prep.* 加上；*adj.* 正的，加的

polyacrylamide *n.* ［化］聚丙烯酰胺

popularity *n.* 普及，流行，声望

positive *adj.* 正的，阳的

potassium *n.* ［化］钾（元素符号K）

potassium bromide 溴化钾（KBr）

potassium hydrogen phthalate 邻苯二甲酸氢钾

potassium permanganate ［化］高锰酸钾

potentiometry 电势［位］测定法

precipitating agent 沉淀剂

precipitation *n.* 沉淀

precision *n.* 精确，精密度，精度

preconcentration *n.* 预选，预富集，预浓缩

precondition *v.* 预处理

predominate *vt.* 掌握，控制，支配；*vi.* 统治，成为主流，支配，占优势

preservation *n.* 保存

pretreatment *n.* 预处理；*adj.* 预处理期间的

previously *adv.* 先前，以前

primary standard 基准物，原标准器

principle *n.* 法则，原则，原理

produce *vt.* 生成，生产，制造，结（果实）

profile *n.* 剖面，侧面，外形，轮廓

programmable *adj.* 可设计的，可编程的

proportional *adj.* 成比例的，相称的，均衡的

proteinaceous *adj.* 蛋白质的，似蛋白质的

proton *n.* ［核］质子

purple *adj.* 紫色的；*n.* 紫色

Q

quantitative *adj.* 定量的，数量的

quantitatively *adv.* 数量上

quartz *n.* 石英

R

ratio *n.* 比，比率

reagent *n.* 反应力，反应物，试剂

receiving flask 接收瓶，收集瓶

recommend *vt.* 推荐，介绍，劝告

rectangle *n.* 长方形，矩形

reddish *adj.* 微红的，略带红色的

redox *n.* 氧化还原作用

reduction *n.* ［化］还原

reference *n.* 参考，参比

reference cell 参比池

reference mark ［测］基准标记

reflux *n.* 回流，逆流，退潮

resistance *n.* 反抗，抵抗，阻抗

resonance *n.* 共振

retention time 保留时间

rhythmic movement 韵律活动，律动

ribbed *adj.* 有肋骨的，有棱纹的

rinse *v.* （用清水）刷，冲洗掉，漂净

rocking *adj.* 摇摆的，摇动的

room temperature 室温，常温（约 20℃）
rotational *adj.* 转动的，轮流的
rub *v.* 擦，摩擦

S

salicylate *n.* ［化］水杨酸盐
sample preparation 样品制备
sample pretreatment 试样预处理
sampling *v.* 取样
saturation *n.* 饱和度
scale *n.* 刻度，衡量，比例
scanning electron microscopy (SEM) 扫描电子显微镜
scatter *v.* 分散，散开，撒开，驱散
scissor *vt.* 剪，剪取；*n.* 剪刀
scuff *vi.* 磨损；*vt.* 使磨损
selenium *n.* ［化］硒（元素符号 Se）
self-check 自检
semipermeable *adj.* 半透性的
sensitivity *n.* 敏感，灵敏（度），灵敏性
sensor *n.* 传感器
septum *n.* 隔膜
shear *v.* 剪，修剪，剪切
shield *n.* 防护物，护罩，盾，盾状物
silanize *vt.* 使硅烷化
silicon *n.* ［化］硅（元素符号 Si）
silver nitrate ［化］硝酸银
similar *adj.* 相似的，类似的
slot *n.* 缝，狭槽
sludge *n.* 软泥，淤泥，矿泥，煤泥
soap *n.* 肥皂
sodium hydroxide ［化］氢氧化钠
software *n.* 软件
solidify *v.* （使）凝固，（使）团结，巩固
solution *n.* 溶液
sort *n.* 种类，类别；*v.* 分类，拣选

specify *vt.* 指定，详细说明，列入清单
spectrometer *n.* ［物］分光计
spectrophotometer *n.* 分光光度计
spectroscopic *adj.* 分光镜的
spectroscopy *n.* ［物］光谱学，波谱学，分光镜使用
spectrum *n.* 光，光谱，型谱
spill *n.* 溢出，溅出；*vt.* 使溢出
spin *v.* 旋转；*n.* 旋转
squeeze *n.* 压榨，挤；*v.* 压榨，挤，挤榨
stable *adj.* 稳定的
stainless steel 不锈钢
stand-alone *n.* 单机
standard *n.* 标准，规格，本位；*adj.* 标准的，第一流的
stationary phase （色谱分析用的）固定相
stirrer *n.* 搅拌器，搅拌者，搅拌用勺子
stoichiometry *n.* 化学计算（法），化学计量学
stopcock *n.* ［机］活塞，活栓，旋塞阀
stream *n.* 流，一股，一串
stretch *v.* 伸展，伸长；*n.* 伸展
subsequent *adj.* 后来的，并发的
substance *n.* 物质，实质，主旨
suction *n.* 抽气机，抽水泵，吸引
sulfate *n.* ［化］硫酸盐
sulfuric acid ［化］硫酸
suspended solid 悬浮固体
swiftly *adv.* 很快地，即刻
symmetric *adj.* 对称的，均衡的
synthesis *n.* 综合，合成
synthetic *adj.* 合成的

T

tap water 自来水，非蒸馏水
teflon ［化］聚四氟乙烯（塑料，绝缘

材料)
theoretical adj. 理论的
thermal analysis 热分析
Thermal Conductivity Detector (TCD) 热导检测器
thermal modulated differential scanning calorimetric (TMDSC) 热调制差示扫描量热法
thermodynamics n. 热力学
thoroughly adv. 十分地,彻底地
thymol blue 百里酚蓝
tip n. 顶,尖端
tissue n. 薄的纱织品,薄纸,棉纸
titrant n. [化] 滴定剂,滴定(用)标准溶液
titration error 滴定误差
titrimetric adj. [化] 滴定(测量)的
titrimetry n. [化] 滴定分析
topographic adj. 地形学上的
torr n. 托(真空度单位)
toxic adj. 有毒的,中毒的
transparent adj. 透明的,显然的,明晰的
trichlorofluoromethane 三氯氟甲烷(CCl_3F)
tube n. 管,管子
tubular adj. 管状的
tutorial n. 指南

U

ultraviolet adj. 紫外线的,紫外的;n. 紫外线辐射
universal indicator 通用指示剂
unreduced adj. 未减少(或缩小、减短)的

V

valence n. [化] (化合)价,原子价

vaporize v. (使)蒸发
verify vt. 检验,校验,查证,核实
vertical adj. 垂直的,直立的
vibration n. 振动,颤动,摇动,摆动
vice versa 反之亦然
violet n. 紫罗兰;adj. 紫罗兰色的
viscosity n. 黏质,黏性
visible spectrum 可见光谱
volatile adj. 飞行的,挥发性的;n. 挥发物
Volhard method 福尔哈德法
voltage n. 电压,伏特数
voltmeter n. 伏特计
volume n. 体积
volumetric flask (容)量瓶

W

wag vt. 摇摆,摇动,饶舌
wash bottle 洗瓶
water tap 水龙头
wavelength n. 波长
whack vt. 重打,击败;vi. 重击
white reference 基准白(色)
width n. 宽度,宽广
wipe v. 擦,揩,擦去
wire n. 金属丝,电线
wire gauze 金属细网纱
wool n. 羊毛,毛织品,毛线,绒线,毛料衣物
wrench n. 扳钳,扳手

Z

Zeeman effect 塞曼效应
zinc n. 锌(元素符号Zn)

Appendix II

A Table of Chemical Elements

化学元素表

元素名称 中文	元素名称 英文	符号	原子序数	元素名称 中文	元素名称 英文	符号	原子序数
氢	hydrogen	H	1	锝	technetium	Tc	43
氦	helium	He	2	钌	ruthenium	Ru	44
锂	lithium	Li	3	铑	rhodium	Rh	45
铍	beryllium	Be	4	钯	palladium	Pd	46
硼	boron	B	5	银	silver	Ag	47
碳	carbon	C	6	镉	cadmium	Cd	48
氮	nitrogen	N	7	铟	indium	In	49
氧	oxygen	O	8	锡	tin	Sn	50
氟	fluorine	F	9	锑	antimony	Sb	51
氖	neon	Ne	10	碲	tellurium	Te	52
钠	sodium	Na	11	碘	iodine	I	53
镁	magnesium	Mg	12	氙	xenon	Xe	54
铝	aluminium	Al	13	铯	cesium	Cs	55
硅	silicon	Si	14	钡	barium	Ba	56
磷	phosphorus	P	15	镧	lanthanum	La	57
硫	sulphur	S	16	铈	cerium	Ce	58
氯	chlorine	Cl	17	镨	praseodymium	Pr	59
氩	argon	Ar	18	钕	neodymium	Nd	60
钾	potassium	K	19	钷	promethium	Pm	61
钙	calcium	Ca	20	钐	samarium	Sm	62
钪	scandium	Sc	21	铕	europium	Eu	63
钛	titanium	Ti	22	钆	gadolinium	Gd	64
钒	vanadium	V	23	铽	terbium	Tb	65
铬	chromium	Cr	24	镝	dysprosium	Dy	66
锰	manganese	Mn	25	钬	holmium	Ho	67
铁	iron	Fe	26	铒	erbium	Er	68
钴	cobalt	Co	27	铥	thulium	Tm	69
镍	nickel	Ni	28	镱	ytterbium	Yb	70
铜	copper	Cu	29	镥	lutetium	Lu	71
锌	zinc	Zn	30	铪	hafnium	Hf	72
镓	gallium	Ga	31	钽	tantalum	Ta	73
锗	germanium	Ge	32	钨	tungsten	W	74
砷	arsenic	As	33	铼	rhenium	Re	75
硒	selenium	Se	34	锇	osmium	Os	76
溴	bromine	Br	35	铱	iridium	Ir	77
氪	krypton	Kr	36	铂	platinum	Pt	78
铷	rubidium	Rb	37	金	gold	Au	79
锶	strontium	Sr	38	汞	mercury	Hg	80
钇	yttrium	Y	39	铊	thallium	Tl	81
锆	zirconium	Zr	40	铅	lead	Pb	82
铌	niobium	Nb	41	铋	bismuth	Bi	83
钼	molybdenum	Mo	42	钋	polonium	Po	84

续表

元素名称		符 号	原子序数	元素名称		符 号	原子序数
中文	英文			中文	英文		
砹	astatine	At	85	锿	einsteinium	Es	99
氡	radon	Rn	86	镄	fermium	Fm	100
钫	francium	Fr	87	钔	mendelevium	Md	101
镭	radium	Ra	88	锘	nobelium	No	102
锕	actinium	Ac	89	铹	lawrencium	Lr	103
钍	thorium	Th	90	𬬻	rutherfordium	Rf	104
镤	protactinium	Pa	91	𬭊	dubnium	Db	105
铀	uranium	U	92	𬭳	seaborgium	Sg	106
镎	neptunium	Np	93	𬭛	bohrium	Bh	107
钚	plutonium	Pu	94	𬭶	hassium	Hs	108
镅	americium	Am	95	䥑	meitnerium	Mt	109
锔	curium	Cm	96	𫟼	darmstadium	Ds	110
锫	berkelium	Bk	97	𬬮	roentgenium	Rg	111
锎	californium	Cf	98				